INTRODUCTION TO POWDER SURFACE AREA

INTRODUCTION TO POWDER SURFACE AREA

S. LOWELL

A WILEY-INTERSCIENCE PUBLICATION
JOHN WILEY & SONS, New York
Chichester • Brisbane • Toronto

Library of Congress Cataloging in Publication Data
Lowell, Seymour, 1931–
Introduction to powder surface area.

"A Wiley-Interscience publication."
Bibliography: p.
Includes index.
1. Powders—Surfaces. 2. Surfaces—Areas and volumes. I. Title.

TA418.78.L68 620'.43 79-15878
ISBN 0-471-04771-6.

Printed in the United States of America

10 9 8 7 6 5 4 3 2 1

My deepest and most sincere appreciation is given to

Dr. Stephen Brunauer and Dr. Joan Shields

for their constructive advice, critical evaluation
and innumerable helpful suggestions

For their encouragement, patience, and endurance
I wish to thank my family

Marilyn, Lauren, and Scott

Preface

The rapid growth of interest in powders and their surface properties in many diverse industries prompted the writing of this book for those who have the need to make meaningful measurements without the benefit of years of experience. It is intended as an introduction to some of the elementary theory and experimental methods used to study the surface area and porosity of solid surfaces. It may be found useful by those with little or no training in solid surfaces who have the need to quickly learn the rudiments of surface area and pore size measurements.

Albertson, New York
January 1979

S. LOWELL

Contents

PART ONE, THEORETICAL

Symbols

Use of symbols for purposes other than those indicated in the following table are so defined in the text. Some symbols not shown in the table are also defined in the text.

$\mathscr{A}$	adsorbate cross-sectional area
A	condensation coefficient
C	BET constant
D	diameter
E	Polanyi adsorption potential
f	permeability aspect factor
G	Gibbs free energy
H	enthalpy
ΔH_i	heat of immersion
h_i	heat of immersion per unit area
i	BET intercept
K	Harkins–Jura constant
L	heat of liquefaction
M	adsorbent weight
$\bar{M}$	molecular weight
N	number of molecules
$\bar{N}$	Avogadro's number
N_m	number of molecules in a monolayer
$\mathscr{N}$	molecular collisions per square centimeter per second
n	moles
P	pressure
P_0	saturated vapor pressure
p	porosity
PSIA	pounds per square inch absolute
PSIG	pounds per square inch gauge
R	gas constant
r	radius
s	BET slope
S	weight-specific surface area
S_t	total surface area
ΔS	entropy change
t	statistical depth
T	absolute temperature

V	volume	θ_i	fraction of surface covered by molecules of depth i
$\bar{V}$	molar volume		
v	linear flow velocity	θ	wetting angle
W	adsorbate weight	μ	micrometers (10^{-6} m)
W_m	monolayer weight	ν	vibrational frequency
		π	spreading pressure
β	affinity coefficient	ρ	density
γ	surface tension	τ	monolayer depth, time for one cycle
η	coefficient of viscosity		

PART ONE

THEORETICAL

CHAPTER ONE

Introduction

1. Real Surfaces

There is a convenient mathematical idealization which asserts that a cube of edge length, l cm, possesses a surface area of $6l^2$ cm^2 and that a sphere of radius r cm exhibits $4\pi r^2$ cm^2 of surface. In reality however, mathematical, perfect, or ideal geometric forms are unattainable since under microscopic examinations all real surfaces exhibit flaws. For example, if a "super microscope" were available one would observe surface roughness not only due to the atomic or molecular orbitals at the surface but also due to voids, steps, pores, and other surface imperfections. These surface imperfections will always create real surface area greater than the corresponding geometric area.

2. Factors Affecting Surface Area

When a cube, real or imaginary, of one meter edge length is subdivided into smaller cubes each one micrometer (10^{-6} meter) in length there will be formed 10^{18} particles each exposing an area of 6×10^{-12} square meters (m^2). Thus, the total area of all the particles is 6×10^6 m^2. This millionfold increase in exposed area is typical of the large surface areas exhibited by fine powders when compared to undivided material. Whenever matter is divided into smaller particles new surfaces must be produced with a corresponding increase in surface area.

In addition to particle size, the particle shape contributes to the powder's surface area. Of all geometric forms, a sphere exhibits the minimum area-to-volume ratio while a chain of atoms, bonded only along the chain axis, will give the maximum area-to-volume ratio. All particulate matter possess geometries and therefore surface areas between these two extremes. The dependence of surface area upon particle shape is readily shown by considering two particles of equal weight M, one particle a cube of edge length l and the other spherical with radius r. Since the particle density ρ is independent of particle shape one can write

$$M_{\text{cube}} = M_{\text{sphere}} \tag{1.1}$$

$$(V\rho)_{\text{cube}} = (V\rho)_{\text{sphere}} \tag{1.2}$$

$$l^3_{\text{cube}} = \tfrac{4}{3}\pi r^3_{\text{sphere}} \tag{1.3}$$

$$\frac{S_{\text{cube}} l_{\text{cube}}}{6} = S_{\text{sphere}} \frac{r_{\text{sphere}}}{3} \tag{1.4}$$

$$\frac{S_{\text{cube}}}{S_{\text{sphere}}} = \frac{2r_{\text{sphere}}}{l_{\text{cube}}} \tag{1.5}$$

Thus, for particles of equal weight, the cubic area, S_{cube}, will exceed the spherical area S_{sphere}, by a factor of $2r/l$.

The range of specific area* can vary widely depending upon the particles size and shape and also the porosity.† The influence of pores can often overwhelm the size and external shape factors. For example, a powder consisting of spherical particles exhibits a total surface area, S_t, as described by equation [1.6]:

$$S_t = 4\pi(r_1^2 N_1 + r_2^2 N_2 + \cdots + r_i^2 N_i) = 4\pi \sum_{i=1} r_i^2 N_i \tag{1.6}$$

where r_i and N_i are the average radii and number of particles, respectively, in the size range i. The volume of the same powder

*The area exposed by one gram of powder is called the "specific surface area".
†Porosity is defined here as surface flaws that are deeper than they are wide.

sample is

$$V = \tfrac{4}{3}\pi(r_1^3 N_1 + r_2^3 N_2 + \cdots + r_i^3 N_i) = \tfrac{4}{3}\pi \sum_{i=1} r_i^3 N_i \qquad [1.7]$$

Replacing V in equation [1.7] by the ratio of mass to density, M/ρ, and dividing equation [1.6] by [1.7] gives the specific surface area

$$S = \frac{S_t}{M} = \frac{3}{\rho} \frac{\Sigma_i N_i r_i^2}{\Sigma_i N_i r_i^3} \qquad [1.8]$$

For spheres of uniform radius equation [1.8] becomes

$$S = \frac{3}{\rho r} \qquad [1.9]$$

Thus, powders with densities near 3 g cm^{-3} will exhibit surface areas about 10^5 cm^2 g^{-1} (10 m^2 g^{-1}) if the particles are spherical with 0.1 micrometer (μm) radius. With radii of 1.0 μm the surface area would decrease tenfold to about 1.0 m^2 g^{-1}. However, if the same 1.0 μm radii particles contained extensive porosity, they could exhibit specific surface areas well in excess of 1000 m^2 g^{-1}, clearly indicating the significant contribution that pores can make to the surface area.

CHAPTER TWO

Surface Area from Size Distributions

Although particulates can assume all regular geometric shapes and in most instances highly irregular shapes, most particle size measurements are based on the so-called "equivalent spherical diameter." This is the diameter of a sphere which would behave in the same manner as the test particle being measured in the same instrument. For example, the Coulter counter[1] is a commonly used instrument for determining particle sizes. Its operation is based on the momentary increase in the resistance of an electrolyte solution which results when a particle passes through a narrow aperture between two electrodes. The resistance change is registered in the electronics as a rapid pulse. The pulse height is proportional to the particle volume, and therefore the particles are sized as equivalent spheres.

Stokes' law[2] is another concept around which several instruments are designed to give particle size or size distributions. Stokes' law is used to determine the settling velocity of particles in a fluid medium as a function of their size. Equation [2.1] is a useful form of Stokes' law

$$D=\sqrt{\frac{18\eta v}{(\rho_s-\rho_m)g}} \qquad [2.1]$$

where D is the particle diameter, η is the coefficient of viscosity, v is the settling velocity, g is the gravitational constant, and ρ_s and ρ_m are the densities of the solid and the medium, respectively. Allen[3] gives an excellent discussion of the various experimental methods associated with sedimentation size analysis. Regardless of the experimental method employed, nonspherical particles will be measured as larger or smaller equivalent spheres depending on whether the particles settle faster or more slowly than spheres of the same mass.

Modifications of Stokes' law have been used in centrifugal devices to enhance the settling rates but are subject to the same limitations of yielding only the equivalent spherical diameter.

Optical devices based upon particle attenuation of a light beam or measurement of scattering angles also give equivalent spherical diameters.

Permeametric methods, discussed in a later chapter, are often used to determine average particle size. The method is based upon the impedance offered to the fluid flow by a packed bed of powder. Again, equivalent spherical diameter is the calculated size.

Sieving is another technique, which sizes particles according to their smallest dimension but gives no information on particle shape.

Electron microscopy can be used to estimate a particles shape, at least in two dimensions, and is further limited to viewing relatively few particles.

Attempts to measure surface area based on any of the preceding methods will give results significantly less than the true value in some cases by factors of 10^3 or greater depending upon the particles shape, surface irregularities, and porosity. At best, surface areas calculated from particle size will establish the lower limit by the implicit assumptions of sphericity or some other regular geometric shape, and by ignoring the highly irregular nature of real surfaces.

CHAPTER THREE

Gas Adsorption

1. Introduction

Examination of powdered materials with an electron microscope can generally disclose the presence of surface imperfections and pores. However, these imperfections or irregularities smaller than the microscope's resolving power will remain hidden. Also hidden from view are the internal structure of the pores, their inner shape and dimensions, their volume and volume distribution, and their contribution to the surface area. However, by enveloping each particle of a powder sample in an adsorbed film, the method of gas adsorption can probe the surface irregularities and pore interiors even at the atomic level. In this manner a very powerful method is available which can generate detailed information about the morphology of surfaces.

To some extent adsorption always occurs when a clean solid surface is exposed to vapor.* Invariably the amount adsorbed on a solid surface will depend upon the absolute temperature T, the pressure P, and the interaction potential E between the vapor (adsorbate) and the surface (adsorbent). Therefore, at some equi-

*A vapor is defined here as a gas below its critical temperature and, therefore, condensable.

librium pressure and temperature the weight W of gas adsorbed on a unit weight of adsorbent is given by

$$W = F(P, T, E) \qquad [3.1]$$

Usually the quantity adsorbed is measured at constant temperature and equation [3.1] reduces to

$$W = F(P, E) \qquad [3.2]$$

A plot of W versus P, at constant T, is referred to as the adsorption isotherm of a particular vapor–solid interface. Were it not for the fact that E, the interaction potential, varies with the properties of the vapor and the solid and also changes with the extent of adsorption, all adsorption isotherms would be identical.

2. Physical and Chemical Adsorption

Depending upon the strength of the interaction, all adsorption processes can be divided into the two categories of chemical or physical adsorption. The former, also called irreversible or chemisorption, is characterized mainly by large interaction potentials, which lead to high heats of adsorption often approaching the value of chemical bonds. This fact, coupled with other spectroscopic, electron spin resonance, and magnetic susceptibility measurements confirms that chemisorption involves true chemical bonding of the gas or vapor with the surface. Because chemisorption occurs through chemical bonding it is often found to occur at temperatures above the adsorbates critical temperature. Strong bonding to the surface is necessary, in the presence of higher thermal energies, if adsorption is to occur at all. Also, as is true for most chemical reactions, chemisorption is usually associated with an activation energy. In addition, chemisorption is necessarily restricted to, at most, a single layer of chemically bound adsorbate at the surface. Another important factor relating to chemisorption is that the adsorbed molecules are localized on the surface. Because of the formation of a chemical bond between an adsorbate molecule and a specific site on the surface the adsorbate is not free to migrate about the surface. This fact often enables the number of active sites on catalysts to be determined by simply measuring the quantity of chemisorbed gas.

The second category, reversible or physical adsorption, exhibits characteristics that makes it most suitable for surface area determinations as indicated by the following:

1. Physical adsorption is accompanied by low heats of adsorption with no violent or disruptive structural changes occurring to the surface during the adsorption measurement.
2. Unlike chemisorption, physical adsorption may lead to surface coverage by more than one layer of adsorbate. Thus, pores can be filled by the adsorbate for pore volume measurements.
3. At elevated temperatures physical adsorption does not occur or is sufficiently slight that relatively clean surfaces can be prepared on which to make accurate surface area measurements.
4. Physical adsorption equilibrium is achieved rapidly since no activation energy is required as in chemisorption. An exception here is adsorption in small pores where diffusion can limit the adsorption rate.
5. Physical adsorption is fully reversible, enabling both the adsorption and desorption processes to be studied.
6. Physically adsorbed molecules are not restrained to specific sites and are free to cover the entire surface. For this reason surface areas rather than number of sites can be calculated.

3. Physical Adsorption Forces

Upon adsorption, the entropy change of the adsorbate, ΔS_a, is necessarily negative since the condensed state is more ordered than the gaseous state because of the loss of at least one degree of translational freedom. A reasonable assumption for physical adsorption is that the adsorbent's entropy remains essentially constant and certainly does not increase by more than the adsorbate's decrease. Therefore, ΔS for the entire system is necessarily negative. The spontaneity of the adsorption process requires that the Gibbs free energy, ΔG, also be a negative quantity. Based upon the entropy and free energy changes, the enthalpy change, ΔH, accompanying physical adsorption is always exothermic:

$$\Delta H = \Delta G + T\Delta S \qquad [3.3]$$

An important interaction at the gas–solid interface during physical adsorption is due to dispersion forces. These forces are present regardless of the nature of other interactions and often account for the major part of the adsorbate–adsorbent potential. The nature of dispersion forces was first recognized in 1930 by F. London,[4] who postulated that the electron motion in an atom or molecule would lead to a rapidly oscillating dipole moment. At any instant the lack of symmetry of the electron distribution about the nuclei imparts a transient dipole moment to an atom or molecule, which vanishes when averaged over a longer time interval. When in close proximity, the rapidly oscillating dipoles of neighboring molecules couple into phase with each other leading to a net attracting potential. This phenomenon is associated with the molecular dispersion of light due to the light's electromagnetic field interaction with the oscillating dipole.

Among other adsorbate–adsorbent interactions contributing to adsorption are:

1. **Ion-dipole**—an ionic solid and electrically neutral but polar adsorbate.
2. **Ion-induced dipole**—a polar solid and polarizable adsorbate.
3. **Dipole–dipole**—a polar solid and polar adsorbate.
4. **Quadrapole interactions**—Symmetrical molecules with atoms of different electronegativities such as CO_2 possess no dipole moment but do have a quadrapole ($^{-}O{-}^{+}C^{+}{-}O^{-}$), which can lead to interactions with polar surfaces.

From the aforesaid it is evident that adsorption forces are similar in nature and origin to the forces that lead to liquefaction of vapors and that the same intermolecular interactions are responsible for both phenomena. Thus, those vapors with high boiling points and, therefore, strong intermolecular interactions will also tend to be strongly adsorbed.

Above the critical temperature the thermal energy possessed by gas molecules is sufficient to overcome the forces leading to liquefaction. Because of the similarity between adsorption and liquefaction, the critical temperature can be used as an estimate of the maximum temperature at which significantly measurable amounts of physical adsorption can occur.

CHAPTER FOUR

Adsorption Isotherms

Brunauer, Deming, Deming and Teller,[5] based upon an extensive literature survey, found that all adsorption isotherms fit into one of the five types shown in Figure 4-1.

The five isotherm shapes depicted in Figure 4-1 each reflects some unique condition. Each of these five isotherms and the conditions leading to its occurrence are discussed in this chapter.

Type I isotherms are encountered when adsorption is limited to, at most, only a few molecular layers. This condition is encountered in chemisorption where the asymptotic approach to a limiting quantity indicates that all of the surface sites are occupied. In the case of physical adsorption, type I isotherms are encountered with microporous powders whose pore size does not exceed a few adsorbate molecular diameters. A gas molecule, when inside pores of these small dimensions, encounters the overlapping potential from the pore walls which enhances the quantity adsorbed at low relative pressures. At higher pressures the pores are filled by adsorbed or condensed adsorbate leading to the plateau showing little or no additional adsorption after the micropores have been filled. Physical adsorption that produces the type I isotherm indicates that the pores are microporous and that the exposed surface resides almost exclusively within the micropores which, once filled with adsorbate, leave little or no external surface for additional adsorption.

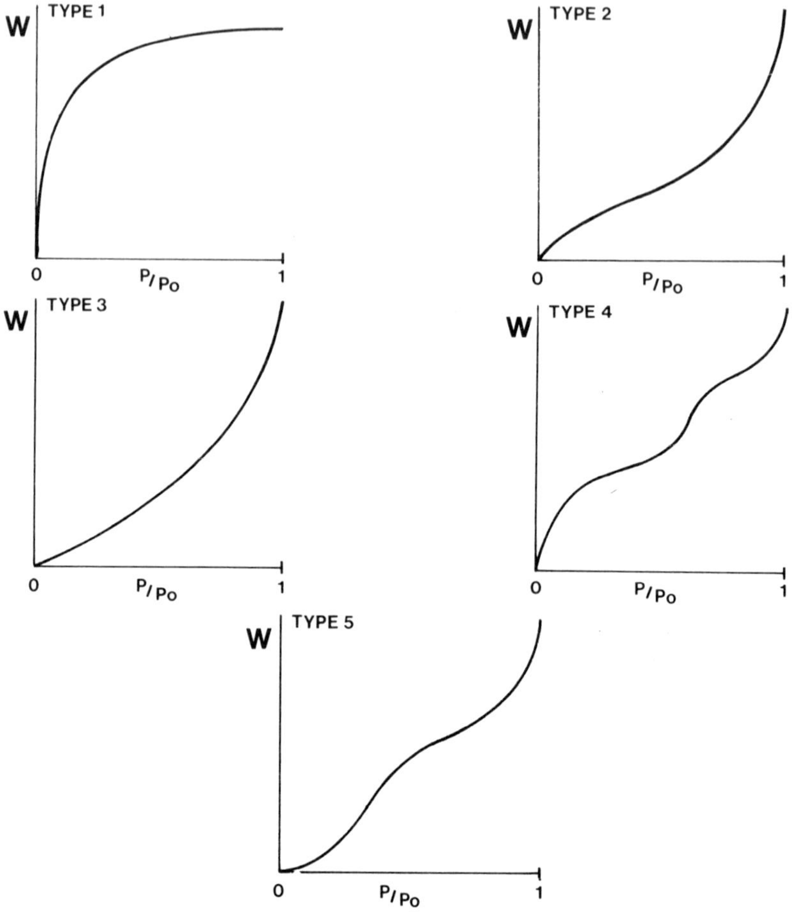

Figure 4-1 The five isotherm classifications according to BDDT. W, weight adsorbed; P, adsorbate equilibrium pressure; P_0, adsorbate saturated equilibrium vapor pressure; P/P_0, relative pressure. Condensation occurs at $P/P_0 \geqslant 1$.

Type II isotherms are most frequently encountered when adsorption occurs on nonporous powders or on powders with pore diameters larger than micropores. The inflection point or knee of the isotherm usually occurs near the completion of the first adsorbed monolayer and, with increasing relative pressure, second and higher layers are completed until at saturation the number of adsorbed layers becomes infinite.

Type III isotherms are characterized principally by heats of adsorption that are less than the adsorbate heat of liquefaction. Thus, as adsorption proceeds, additional adsorption is facilitated because the adsorbate interaction with an adsorbed layer is greater than the interaction with the adsorbent surface.

Type IV isotherms occur on porous adsorbents possessing pores in the radius range of approximately 15–1000 angstroms (Å). The slope increase at higher elevated pressures indicates an increased uptake of adsorbate as the pores are being filled. As is true for the type II isotherms, the knee of the type IV isotherm generally occurs near the completion of the first monolayer.

Type V isotherms result from small adsorbate–adsorbent interaction potentials similar to the type III isotherms. However, type V isotherms are also associated with pores in the same range as those of the type IV isotherms.

CHAPTER FIVE

Langmuir and BET Theories (Kinetic Isotherms)

The success of kinetic theories directed toward the measurements of surface areas depends upon their ability to predict the number of adsorbate molecules required to exactly cover the solid with a single molecular layer. Equally important is the cross-sectional area of each molecule or the effective area covered by each adsorbed molecule on the surface. The surface area then, is the product of the number of molecules in a completed monolayer times the effective cross-sectional area of an adsorbate molecule. The number of molecules required for the completion of a monolayer is considered in this chapter while Chapter 6 discusses the adsorbate cross-sectional area.

1. The Langmuir Isotherm, Type I

The asymptotic approach of the quantity adsorbed toward a limiting value indicates that type I isotherms are limited to, at most, a few monolayers.* In the case of chemisorption, only one layer can be

*Physical adsorption on microporous materials show type I isotherms because the pores limit adsorption to only a few molecular layers. Once the micropores are filled there is only a small fraction of the original surface area exposed for continued adsorption.

bonded to the surface and, therefore, chemisorption always exhibits a type I isotherm.† Although it is possible to calculate the number of molecules in the monolayer from the type I chemisorption isotherm, some serious difficulty is encountered when attempts are made to apply the cross-sectional adsorbate area. This difficulty arises because chemisorption tightly binds and localizes the adsorbate to a specific surface site so that the spacing between adsorbed molecules will depend upon the adsorbent surface structure as well as the size of the adsorbed molecules or atoms. In those cases where the surface sites are widely separated, the calculated surface area will be smaller than the actual value because the number of molecules in the monolayer will be less than the maximum number that the surface can accommodate. Nevertheless, it is instructive to consider the type I isotherm in preparation for the more rigorous requirements of the other four types.

Using a kinetic approach, Langmuir was able to describe the type I isotherm with the assumption that adsorption was limited to a monolayer. According to the kinetic theory of gases, the number of molecules $\mathcal{N}$ striking each square centimeter of surface per second is given by

$$\mathcal{N} = \frac{\overline{N}P}{(2\pi \overline{M}RT)^{1/2}} \qquad [5.1]$$

where $\overline{N}$ is Avogadro's number, P is the adsorbate pressure, $\overline{M}$ is the adsorbate molecular weight, R is the gas constant, and T is in degrees Kelvin. If θ_0 is the fraction of the surface unoccupied (i.e., with no adsorbed molecules) then the number of collisions with bare or uncovered surface per square centimeter of surface each second is written as

$$\mathcal{N}' = kP\theta_0 \qquad [5.2]$$

The constant k is $\overline{N}/(2\pi\overline{M}RT)^{1/2}$. The number of molecules striking and adhering to each square centimeter of surface is

$$\mathcal{N}_{\text{ads}} = kP\theta_0 A_1 \qquad [5.3]$$

†Under certain conditions physical adsorption can occur on top of a chemisorbed layer, but this does not change the essential point being made here.

the term A_1 is the condensation coefficient and represents the probability of a molecule being adsorbed upon collision with the surface.

The rate at which adsorbed molecules leave each square centimeter of surface is given by

$$\mathcal{N}_{\text{des}} = N_m \theta_1 v_1 e^{-E/RT} \qquad [5.4]$$

where N_m is the number of adsorbate molecules in a completed monolayer of one square centimeter, θ_1 is the fraction of the surface occupied by the adsorbed molecules, E is the energy of adsorption, and v_1 is the adsorbate's vibrational frequency normal to the surface when adsorbed. Actually, the product $N_m\theta_1$ is the number of molecules adsorbed per square centimeter. Multiplication by v_1 converts this number of molecules to the maximum rate at which they can leave the surface. The term $e^{-E/RT}$ represents the probability of an adsorbed molecule possessing adequate energy to overcome the net attractive potential of the surface. Thus, equation [5.4] contains all the parameters required to describe the rate at which molecules leave each square centimeter of surface.

At equilibrium the rates of adsorption and desorption are equal. From equations [5.3] and [5.4] one obtains

$$N_m \theta_1 v_1 e^{-E/RT} = kP\theta_0 A_1 \qquad [5.5]$$

Recognizing that $\theta_0 = 1 - \theta_1$, yields

$$N_m \theta_1 v_1 e^{-E/RT} = kPA_1 - \theta_1 kPA_1 \qquad [5.6]$$

then

$$\theta_1 = \frac{kPA_1}{N_m v_1 e^{-E/RT} + kPA_1} \qquad [5.7]$$

allowing

$$K = \frac{kA_1}{N_m v_1 e^{-E/RT}} \qquad [5.8]$$

Substituting equation [5.8] into [5.7] gives

$$\theta_1 = \frac{KP}{1+KP} \qquad [5.9]$$

The assumption implicit in equation [5.8] is that the adsorption energy E is constant, which implies an energetically uniform surface.

Up to and including one layer of coverage one can write

$$\theta_1 = \frac{N}{N_m} = \frac{W}{W_m} \qquad [5.10]$$

where N and N_m are the number of molecules in the incompleted and completed monolayer, respectively, and W/W_m is the weight adsorbed relative to the weight adsorbed in a completed monolayer. Substituting W/W_m for θ_1 in equation (5.9) yields

$$\frac{W}{W_m} = \frac{KP}{1+KP} \qquad [5.11]$$

Equation [5.11] is the Langmuir equation for type I isotherms. Rearrangement of equation [5.11] gives

$$\frac{P}{W} = \frac{1}{KW_m} + \frac{P}{W_m} \qquad [5.12]$$

A plot of P/W versus P will give a straight line of slope $1/W_m$ and intercept $1/KW_m$ from which both K and W_m can be calculated.

Having established W_m, the sample surface area S_t can be calculated from equation [5.13]

$$S_t = N_m \mathscr{A} = \frac{W_m \overline{N} \mathscr{A}}{\overline{M}} \qquad [5.13]$$

where $\mathscr{A}$ is the cross-sectional area, $\overline{M}$ is the molecular weight of an adsorbate molecule, and $\overline{N}$ is Avogadro's number.

Although the Langmuir equation describes chemisorption and type I isotherms, it fails to be adequately general to treat physical adsorption and the type II–type V isotherms. In addition, surface area measurements made from type I isotherms are subject to uncertainties regardless of whether chemisorption or physical adsorption is occurring. With chemisorption, localization of the adsorbate molecules leaves the value of $\mathscr{A}$ seriously in question since the adsorbate will adsorb only at active surface sites leaving an unspecified area around each chemisorbed molecule. When applied to physical adsorption, the type I isotherm is associated with condensation in micropores with no clearly defined region of monolayer coverage.

2. The Brunauer, Emmett, and Teller (BET) Theory[7]

During the process of physical adsorption, at very low relative pressure, the first sites to be covered are the more energetic ones. Those sites with higher energy, on a chemically pure surface, reside within narrow pores where the pore walls provide overlapping potentials. Other high-energy sites lie between the horizontal and vertical edges of surface steps where the adsorbate can interact with surface atoms in two planes. In general, wherever the adsorbate is afforded the opportunity to interact with overlapping potentials or an increased number of surface atoms, there will be a higher-energy site. On surfaces consisting of heteroatoms, such as organic solids or impure materials, there will be variations in adsorption potential depending upon the nature of the atoms or functional groups exposed at the surface.

That the more energetic sites are covered first as the pressure is increased does not mean that no adsorption occurs on sites of less potential. Rather, it implies that the average residence time of a physically adsorbed molecule is longer on the higher-energy sites. Accordingly, as the adsorbate pressure is allowed to increase, the surface becomes progressively coated and the probability increases that a gas molecule will strike and be adsorbed on a previously bound molecule. Clearly then, prior to complete surface coverage there will commence the formation of second and higher adsorbed layers. In reality, there exists no pressure at which the surface is covered with exactly a completed physically adsorbed monolayer. The effectiveness of the Brunauer, Emmett, and Teller (BET) theory is that it enables an experimental determination of the number of molecules required to form a monolayer despite the fact that exactly one monomolecular layer is never actually formed.

Brunauer, Emmett, and Teller, in 1938, extended Langmuir's kinetic theory to multilayer adsorption. The BET theory assumes that the uppermost molecules in adsorbed stacks are in dynamic equilibrium with the vapor. This means that, where the surface is covered with only one layer of adsorbate, an equilibrium exists between that layer and the vapor, and where two layers are adsorbed, the upper layer is in equilibrium with the vapor, and so forth. Since the equilibrium is dynamic the actual location of the surface sites covered by one, two, or more layers may vary, but the number of molecules in each layer will remain constant.

Using the Langmuir theory and equation [5.5] as a starting point to describe the equilibrium between the vapor and the adsorbate in the first layer

$$N_m\theta_1 v_1 e^{-E_1/RT} = kP\theta_0 A_1 \qquad \text{[cf. 5.5]}$$

By analogy, for the fraction of surface covered by two layers one may write

$$N_m\theta_2 v_2 e^{-E_2/RT} = kP\theta_1 A_2 \qquad [5.14]$$

In general, for the *n*th layer one obtains

$$N_m\theta_n v_n e^{-E_n/RT} = kP\theta_{N-1} A_N \qquad [5.15]$$

The BET theory assumes that the terms v, E, and A remain constant for the second and higher layers. This assumption is justifiable only on the grounds that the second and higher layers are all equivalent to the liquid state. This undoubtedly approaches the truth as the layers proceed away from the surface but is somewhat questionable for the layers nearer the surface because of polarizing forces. Nevertheless, using this assumption one can write a series of equations, using L as the heat of liquefaction

$$N_m\theta_1 v_1 e^{-E_1/RT} = kP\theta_0 A_1 \qquad [5.16a]$$

$$N_m\theta_2 v e^{-L/RT} = kP\theta_1 A \qquad [5.16b]$$

$$N_m\theta_3 v e^{-L/RT} = kP\theta_2 A \qquad [5.16c]$$

and in general for the second and higher layers

$$N_m\theta_n v e^{-L/RT} = kP\theta_{n-1} A \qquad [5.16d]$$

From these it follows that

$$\frac{\theta_1}{\theta_0}=\frac{kPA_1}{N_m\nu_1 e^{-E_1/RT}}=\alpha \qquad [5.17a]$$

$$\frac{\theta_2}{\theta_1}=\frac{kPA}{N_m\nu e^{-L/RT}}=\beta \qquad [5.17b]$$

$$\frac{\theta_3}{\theta_2}=\frac{kPA}{N_m\nu e^{-L/RT}}=\beta \qquad [5.17c]$$

$$\frac{\theta_n}{\theta_{n-1}}=\frac{kPA}{N_m\nu e^{-L/RT}}=\beta \qquad [5.17d]$$

then

$$\theta_1=\alpha\theta_0 \qquad [5.18a]$$
$$\theta_2=\beta\theta_1=\alpha\beta\theta_0 \qquad [5.18b]$$
$$\theta_3=\beta\theta_2=\alpha\beta^2\theta_0 \qquad [5.18c]$$
$$\theta_n=\beta\theta_{n-1}=\alpha\beta^{n-1}\theta_0 \qquad [5.18d]$$

The total number of molecules adsorbed at equilibrium is

$$N=N_m\theta_1+2N_m\theta_2+\cdots+nN_m\theta_n=N_m(\theta_1+2\theta_2+\cdots+n\theta_n) \qquad [5.19]$$

Substituting for $\theta_1, \theta_2 \ldots$ from equations [5.18*a*–*d*] gives

$$\frac{N}{N_m}=\alpha\theta_0+2\alpha\beta\theta_0+3\alpha\beta^2\theta_0+\cdots+n\alpha\beta^{n-1}\theta_0$$
$$=\alpha\theta_0(1+2\beta+3\beta^2+\cdots+n\beta^{n-1}) \qquad [5.20]$$

Since both α and β are assumed to be constants, one can write

$$\alpha=C\beta \qquad [5.21]$$

This defines C by using equation [5.17*a*] and [5.17*b*–*d*] as

$$\frac{A_1\nu_2}{A_2\nu_1}\,e^{(E_1-L)/RT}=C \qquad [5.22]$$

Substituting $C\beta$ for α in equation [5.20] yields

$$\frac{N}{N_m} = C\theta_0(\beta + 2\beta^2 + 3\beta^3 + \cdots + n\beta^n) \qquad [5.23]$$

The preceding summation is just $\beta/(1-\beta)^2$. Therefore,

$$\frac{N}{N_m} = \frac{C\theta_0\beta}{(1-\beta)^2} \qquad [5.24]$$

Necessarily

$$1 = \theta_0 + \theta_1 + \theta_2 + \cdots + \theta_n \qquad [5.25]$$

then

$$\theta_0 = 1 - (\theta_1 + \theta_2 + \cdots + \theta_n) = 1 - \sum_{n=1}^{\infty} \theta_n \qquad [5.26]$$

Substituting equation [5.26] into [5.24] gives

$$\frac{N}{N_m} = \frac{C\beta}{(1-\beta)^2}\left(1 - \sum_{n-1}^{\infty} \theta_n\right) \qquad [5.27]$$

Replacing θ_n in equation [5.27] with $\alpha\beta^{n-1}\theta_0$ from equation [5.18d] yields

$$\frac{N}{N_m} = \frac{C\beta}{(1-\beta)^2}\left(1 - \alpha\theta_0 \sum_{n=1}^{\infty} \beta^{n-1}\right) \qquad [5.28]$$

and introducing $C\beta$ from equation [5.21] in place of α gives

$$\frac{N}{N_m} = \frac{C\beta}{(1-\beta)^2}\left(1 - C\theta_0 \sum_{n=1}^{\infty} \beta_n\right) \qquad [5.29]$$

The summation in equation [5.29] is

$$\sum_{n=1}^{\infty} \beta^n = \beta + \beta^2 + \cdots + \beta^n = \frac{\beta}{1-\beta} \qquad [5.30]$$

Then

$$\frac{N}{N_m} = \frac{C\beta}{(1-\beta)^2}\left(1 - C\theta_0 \frac{\beta}{1-\beta}\right) \qquad [5.31]$$

From equation [5.24] we have

$$\frac{C\beta}{(1-\beta)^2}=\frac{N}{N_m}\frac{1}{\theta_0} \qquad \text{[cf. 5.24]}$$

Then equation [5.31] becomes

$$1=\frac{1}{\theta_0}\left(1-C\theta_0\frac{\beta}{1-\beta}\right) \qquad [5.32]$$

and

$$\theta_0=\frac{1}{1+C\beta/(1-\beta)} \qquad [5.23]$$

Introducing θ_0 from equation [5.33] into [5.24] yields

$$\frac{N}{N_m}=\frac{C\beta}{(1-\beta)(1-\beta+C\beta)} \qquad [5.34]$$

When β equals unity, N/N_m becomes infinite. This can physically occur only when adsorbate condenses on the surface or when $P/P_0=1$.

Rewriting equation [5.17d] for $P=P_0$ gives

$$1=\frac{kAP_0}{N_m v e^{-L/RT}} \qquad [5.35]$$

but

$$\beta=\frac{kAP}{N_m v e^{-L/RT}} \qquad \text{[cf. 5.17}d\text{]}$$

then

$$\beta=\frac{P}{P_0} \qquad [5.36]$$

Introducing this value for β into equation [5.34] gives

$$\frac{N}{N_m}=\frac{C(P/P_0)}{(1-P/P_0)[1-P/P_0+C(P/P_0)]} \qquad [5.37]$$

Recalling that $N/N_m=W/W_m$ (equation 5.10) and rearranging equation [5.37] gives the BET equation in final form,

$$\frac{1}{W[(P_0/P)-1]}=\frac{1}{W_mC}+\frac{C-1}{W_mC}\frac{P}{P_0} \qquad [5.38]$$

If adsorption occurs in pores limiting the number of layers then the summation in equation [5.27] is limited to n and the BET equation takes the form

$$\frac{W}{W_m}=\frac{C}{[(P_0/P)-1]}\frac{[1-(n+1)(P/P_0)^n+n(P/P_0)^{n+1}]}{[1+(C-1)P/P_0-C(P/P_0)^{n+1}]} \qquad [5.39]$$

Equation [5.39] reduces to [5.38] with $n=\infty$ and to the Langmuir equation [5.11] with $n=1$.

3. Surface Areas from the BET Equation

The determination of surface areas from the BET theory is a straightforward application of equation [5.38]. A plot of $1/W(P_0/P-1)$ versus P/P_0, as shown in Figure 5-1, will yield a straight line usually in the range $0.05 \leqslant P/P_0 \leqslant 0.35$.

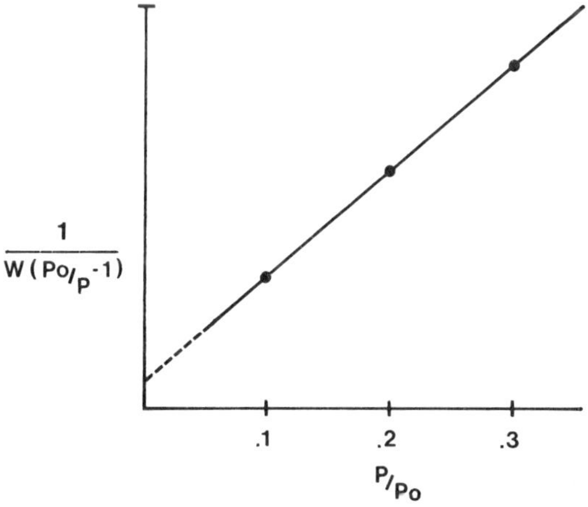

Figure 5-1 Typical BET plot. W, weight adsorbed; W_m, monolayer weight; C, BET constant. The slope and intercept are s and i, respectively.

The slope s and the intercept i of a BET plot are

$$s = \frac{C-1}{W_m C} \qquad [5.40]$$

$$i = \frac{1}{W_m C} \qquad [5.41]$$

Solving the preceding equations for W_m and C gives

$$W_m = \frac{1}{s+i} \qquad [5.42]$$

and

$$C = \frac{s}{i} + 1 \qquad [5.43]$$

The total surface area can be calculated from equation [5.13], viz.,

$$S_t = \frac{W_m \bar{N} \mathscr{A}}{\bar{M}} \qquad [\text{cf. } 5.13]$$

and the specific surface area can be determined by dividing S_t by the sample weight.

4. The Meaning of Monolayer Coverage

Hill[8] has shown that when sufficient adsorption has occurred to cover the surface with exactly one layer of molecules the fraction of surface, θ_0, not covered by any molecules is dependent on the BET C value and is given by

$$\theta_0 = \frac{C^{1/2} - 1}{C - 1} \qquad [5.44]$$

From the preceding equation it is evident that when sufficient adsorption has occurred to form a monolayer there is still always some fraction of surface unoccupied. Indeed, only for C values approaching infinity will θ_0 approach zero and in such cases the high adsorbate–surface interaction can only result from chemisorption. For nominal C values, say near 100, the fraction of surface unoccupied, when exactly sufficient adsorption has occurred to form a monolayer, is 0.091. Therefore, on the average each occupied site contains about 1.1 molecules. The implication here is that the BET equation indicates the weight of adsorbate required to form a single molecular layer on the surface, although no such phenomenon as a uniform monolayer exists in the case of physical adsorption.

5. The BET Constant and Site Occupancy

Equation [5.44] is used to calculate the fraction of surface unoccupied when $W = W_m$, that is, when just a sufficient number of molecules have been adsorbed to give monolayer coverage. Lowell[9] has derived an equation that can be used to calculate the fraction of surface covered by adsorbed molecules of one or more layers in depth. Lowell's equation is

$$(\theta_i)_m = C\left[\frac{C^{1/2}-1}{C-1}\right]^{i+1} \qquad [5.45]$$

where θ_i represents the fraction of surface covered by layers i molecules deep. The subscript m denotes that equation [5.45] is valid only when sufficient adsorption has occurred to make $W = W_m$.

Table 5-1 shows the fraction of surface covered by layers of various depth, as calculated from equations [5.44] and [5.45], as a function of the BET C value.

Equations [5.44] and [5.45] should not be taken to mean that the adsorbate is necessarily arranged in neat stacks of various heights but rather as an indication of the fraction of surface covered with the equivalent of i molecules regardless of their specific arrangement, lateral mobility, and equilibrium with the vapor phase.

Of further interest is the fact that when the BET equation [5.38] is solved for the relative pressure corresponding to monomolecular coverage, ($W = W_m$), one obtains

$$\left(\frac{P}{P_0}\right)_m = \frac{C^{1/2}-1}{C-1} \qquad [5.46]$$

Table 5-1.

	$(\theta_i)_m$			
i	$C=1000$	$C=100$	$C=10$	$C=1$
0	0.0307	0.0909	0.2403	0.5000
1	0.9396	0.8264	0.5772	0.2500
2	0.0288	0.0751	0.1387	0.1250
3	0.0009	0.0068	0.0333	0.0625
4		0.0006	0.0080	0.0313
5		0.0001	0.0019	0.0156
6			0.0005	0.0078
7			0.0001	0.0039
8				0.0019
9				0.0009
10				0.0005
11				0.0002
12				0.0001

For the special case of $C=1$, $(\theta_i)_m$ is evaluated by

$$\lim_{C\to 1}\frac{C^{1/2}-1}{C-1}=0.5$$

The subscript m above refers to monolayer coverage. Equating equations [5.46] and [5.44] produces the interesting fact that

$$\theta_0=\left(\frac{P}{P_0}\right)_m \qquad [5.47]$$

In other words, the numerical value of the relative pressure required to make W equal to W_m is also the fraction of surface unoccupied by adsorbate.

6. Applicability of the BET Theory

Although derived nearly 40 years ago, the BET theory continues to be used by the majority of workers because of its ease of use, its definitiveness, and its ability to accommodate each of the five isotherm types. The mathematical nature of the BET equation in its most general form, [5.39], gives the Langmuir or type I isotherm when $n=1$. Plots of W/W_m versus P/P_0 using equation [5.38] conform to type II or type III isotherms for C values greater than and less than 2, respectively. Figure 5-2 shows the shape of several isotherms for various values of C. The data for Figure 5-2 are shown in Table 5-2 with values of W/W_m calculated from equation [5.38] after rearrangement to

$$\frac{W_m}{W}=\left(1-\frac{P}{P_0}\right)+\frac{1}{C}\left(\frac{P}{P_0}+\frac{P_0}{P}-2\right) \qquad [5.48]$$

The remaining two isotherms, types IV and V, are modifications of the type II and type III isotherms due to the presence of pores.

Rarely, if ever, does the BET theory exactly match an experimental isotherm over its entire range of relative pressures. In a qualitative sense, however, it does provide a theoretical foundation for the various isotherm shapes. Of equal significance is the fact that in the region of relative pressures near completed monolayers ($0.05 \leqslant P/P_0 \leqslant 0.35$) the BET theory and experimental isotherms do agree very well, leading to a powerful and extremely useful method of surface area determination.

The fact that most monolayers are completed in the range $0.05 \leqslant P/P_0 \leqslant 0.35$ reflects the value of most C constants. As shown in Table 5-2, the value of W/W_m equals unity in the previous range of relative pressures for C values between 3 and 1000, which covers the great majority of all isotherms.

The sparsity of data regarding type III isotherms, with C values of 2 or less, leaves open the question of the usefulness of the BET method for determining surface areas when type III isotherms are encountered. Often in this case it is possible to change the adsorbate to one with a higher C value thereby changing the isotherm shape. Brunauer, Copeland, and Kantro,[10] however, point to considerable success in calculating the surface area from type III isotherms as well as predicting the temperature coefficient of the same isotherms.

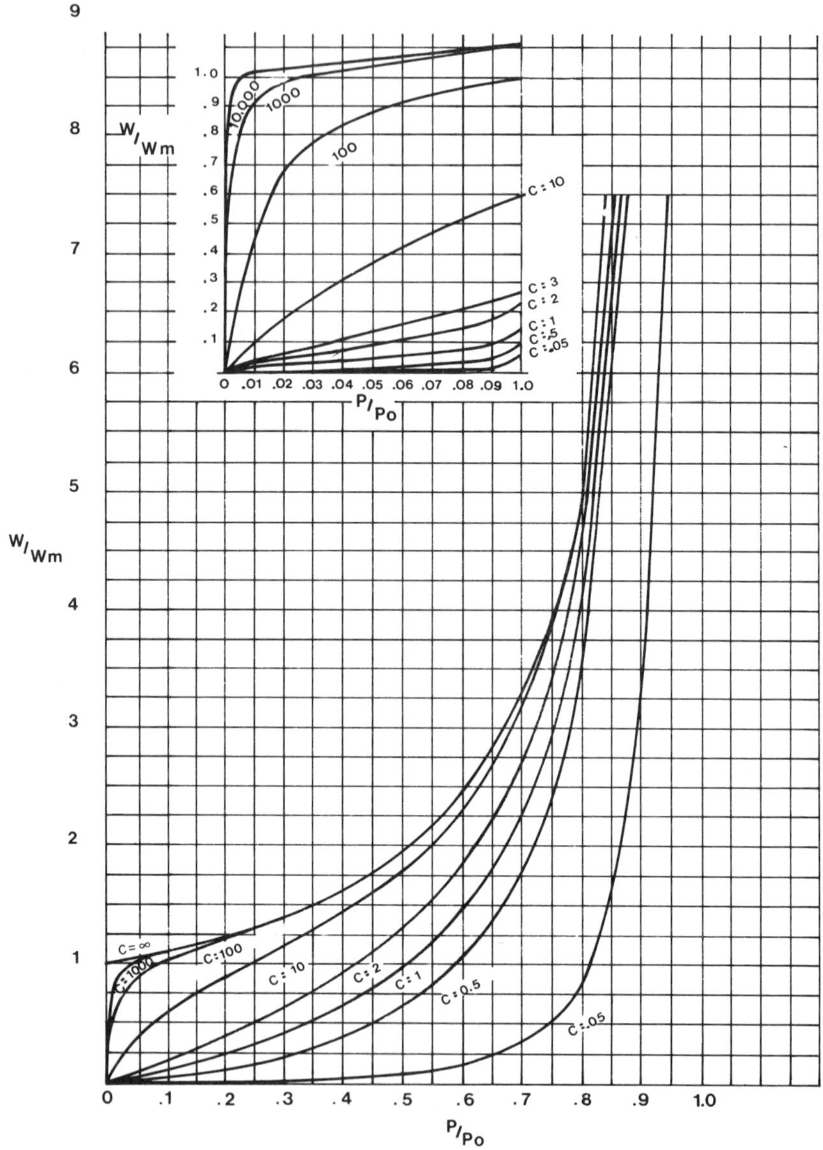

Figure 5-2 Isotherm shapes as a function of BET *C* values.

Table 5-2. Values of W/W , Relative Pressures, and C

				W/W_m							
P/P_0	$1-P/P_0$	P_0/P	$P/P_0+(P_0/P)-2$	$C=0.05$	$C=0.5$	$C=1$	$C=2$	$C=3$	$C=10$	$C=100$	$C=1000$
0.02	0.98	50	48.0	0.001	0.010	0.020	0.040	0.039	0.173	0.685	0.973
0.05	0.95	20	18.1	0.003	0.027	0.053	0.100	0.144	0.363	0.884	1.03
0.10	0.90	10	8.1	0.006	0.058	0.111	0.202	0.278	0.585	1.01	1.10
0.20	0.80	5	3.2	0.015	0.139	0.250	0.417	0.536	0.893	1.20	1.25
0.30	0.70	3.33	1.6	0.030	0.253	0.429	0.660	0.804	1.16	1.40	1.43
0.40	0.60	2.50	0.90	0.054	0.417	0.667	0.952	1.11	1.45	1.64	1.66
0.50	0.50	2.00	0.50	0.095	0.667	1.00	1.33	1.50	1.82	1.98	1.99
0.60	0.40	1.67	0.27	0.172	1.06	1.49	1.87	2.04	2.34	2.48	2.49
0.70	0.30	1.43	0.13	0.345	1.78	2.32	2.74	2.91	3.19	3.32	3.33
0.80	0.20	1.25	0.05	0.833	3.33	4.00	4.44	4.61	4.88	4.99	5.00
0.90	0.10	1.11	0.01	3.33	8.33	9.09	9.52	9.68	9.90	9.99	10.0
0.94	0.06	1.06	0.0038	7.35	14.7	15.67	16.1	16.3	16.5	16.7	16.7

7. Some Criticism of the BET Theory

In spite of the success of the BET theory, some of the assumptions upon which it is founded are not above criticism. One questionable assumption is that of an energetically homogeneous surface, that is, all the adsorption sites are energetically identical. Further, the BET model ignores the influence of lateral adsorbate interactions.

Brunauer[11] answers these criticisms by pointing out that lateral interaction between adsorbate molecules necessarily increases as the surface becomes more completely covered. The adsorption potential, however, decreases with increasing adsorption up to monolayer coverage since on an energetically heterogeneous surface the high-energy sites will be occupied at lower relative pressures with occupancy of the lower-energy sites occurring nearer to completion of the monolayer. Figure 5-3 illustrates that these two trends can lead to a nearly constant overall adsorption energy up to completion of the monolayer, which is an implicit assumption of the BET theory.

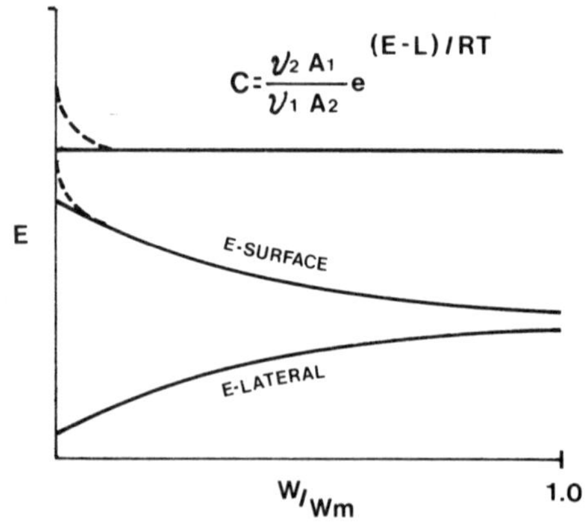

Figure 5-3 Variation in adsorption and lateral interaction potentials. The diagram shows a constant C value which results when the decrease in adsorption potential is matched by the increase in lateral interactions. The dotted lines illustrate the influence of very high adsorption potentials which can account, at least in part, for the failure of BET plots at very low relative pressures ($P/P_0 < 0.05$).

A further criticism of the BET theory is the assumption that the heat of adsorption on the second and higher layers is equal to the heat of liquefaction. It seems reasonable to expect that polarization forces would induce a higher heat of adsorption in the second layer than in the third, and so forth. Only after several layers are adsorbed should the heat of adsorption equal the heat of liquefaction. It is, therefore, difficult to resolve a model of molecules adsorbed in stacks while postulating that all layers above the first are thermodynamically a true liquid structure. The apparent validity of these criticisms contributes to the failure of the BET equation at higher relative pressures ($P/P_0 > 0.35$). However, in the range of relative pressure leading to coverage near $W/W_m = 1$ the BET C values usually give heats of adsorption that are reasonable. Thus, for the great majority of isotherms the range of relative pressures between 0.05 and 0.35, the linear BET range, apparently represents a condition in which the very high energy sites have been occupied and extensive multilayer adsorption has not yet commenced. It is within these limits that the BET theory is generally valid.

Instances have been found in which BET plots were noted to be linear to relative pressures as high as 0.5, and in other cases the linear range is found only below relative pressures of 0.1.[12] The extent to which these deviations from the usual range of linearity reflect unusual surface properties is difficult to ascertain. For example, micropores exhibit unusually high adsorption potentials due to the overlapping potential from the walls of the micropore. Under these enhanced conditions condensation can occur within the pores at relative pressures less than 0.1, and linear BET plots are found at even lower relative pressures.

CHAPTER SIX

The Single Point BET Method[13]

The BET theory requires that a plot of $[W(P_0/P-1)]^{-1}$ versus P/P_0 be linear with a finite intercept [cf. equation [5.38] and Figure 5-1]. By reducing the experimental requirement to only one data point, the single point method offers the advantages of simplicity and speed with often little loss in accuracy.

1. Derivation of the Single Point Method

The slope s and intercept i of a BET plot are

$$s = \frac{C-1}{W_m C} \qquad \text{[cf. 5.40]}$$

$$i = \frac{1}{W_m C} \qquad \text{[cf. 5.41]}$$

Then

$$\frac{s}{i} = C - 1 \qquad [6.1]$$

For reasonably high values of C the intercept is small compared to the slope and may in many instances be taken as zero. With this approximation, equation [5.38], the BET equation, becomes

$$\frac{1}{W(P_0/P-1)}=\frac{C-1}{W_mC}\left(\frac{P}{P_0}\right) \qquad [6.2]$$

since $1/W_mC$, the intercept, is assumed to vanish, equation [6.2] reduces to

$$W_m=W\left(1-\frac{P}{P_0}\right) \qquad [6.3]$$

The total surface area as measured by the single point method is then

$$S_t=W\left(1-\frac{P}{P_0}\right)\frac{\bar{N}}{\bar{M}}\mathscr{A} \qquad [6.4]$$

2. Comparison of the Single Point and Multipoint Methods

The error introduced by the one point method can be evaluated by examining the difference between W_m as determined by equations [6.3] and [5.38], the BET equation. Solving equation [5.38] for W_m gives

$$W_m=W\left(\frac{P_0}{P}-1\right)\left[\frac{1}{C}+\frac{C-1}{C}\left(\frac{P}{P_0}\right)\right] \qquad [6.5]$$

Subtracting equation [6.3] from [6.5] and dividing by equation [6.5] gives the relative error associated with the single point method,

$$\frac{(W_m)_{\mathrm{mp}}-(W_m)_{\mathrm{sp}}}{(W_m)_{\mathrm{mp}}}=\frac{1-P/P_0}{1+(C-1)P/P_0} \qquad [6.6]$$

The subscripts mp and sp, refer to the multi- and single point methods.

Table 6-1 shows the relative error of the single point method compared to the multipoint method as a function of P/P_0 and C as calculated from equation [6.6].

The last column of Table 6-1 is established by substituting equation [5.46] into equation [6.6] for the special case when

Table 6-1. Relative Error Using the Single Point Method at Various Relative Pressures

C	$P/P_0=0.1$	$P/P_0=0.2$	$P/P_0=0.3$	$P/P_0=(P/P_0)_m$[a]
1	0.90	0.80	0.70	0.50
10	0.47	0.29	0.19	0.23
50	0.17	0.07	0.04	0.12
100	0.08	0.04	0.02	0.09
1000	0.009	0.004	0.002	0.03

[a] $(P/P_0)_m$ is the relative pressure that gives monolayer coverage according to a multipoint determination.

$P/P_0=(P/P_0)_m$, thus

$$\frac{(W_m)_{\mathrm{mp}}-(W_m)_{\mathrm{sp}}}{(W_m)_{\mathrm{mp}}}=\frac{C^{1/2}-1}{C-1}=\left(\frac{P}{P_0}\right)_m=\theta_0 \qquad [6.7]$$

The surprising relationship above shows that when a single point analysis is made using the relative pressure which would give monolayer coverage according to the multipoint theory, the relative error will be equal to the relative pressure employed and also, by equation [5.44], will be equal to the fraction of surface unoccupied.

A more explicit insight into the mathematical differences of the multi- and single point methods is obtained by considering a single point analysis using a relative pressure of 0.3 with a corresponding multipoint C value of 100.

From equation [6.3], the single point BET equation, one obtains

$$(W_m)_{\mathrm{sp}}=0.7W_{0.3} \qquad [6.8]$$

The term $(W_m)_{\mathrm{sp}}$ refers to the monolayer weight as determined by the single point method and $W_{0.3}$ is the experimental weight adsorbed at the relative pressure of 0.3.

From equation [5.46] the relative pressure required for monolayer coverage is;

$$\left(\frac{P}{P_0}\right)_m=\frac{100^{1/2}-1}{100-1}=0.0909 \qquad [6.9]$$

Using equation [6.5], the multipoint equation, to find W_m, gives

$$W_m = W_{0.3}(3.33-1)\left[\frac{1}{100}+\frac{100-1}{100}(0.3)\right]=0.716W_{0.3} \qquad [6.10]$$

Comparison of equations [6.10] and [6.8] shows that the difference between the single and multipoint methods is identical to that shown in Table 6-1 for $P/P_0=0.3$ and $C=100$; viz.,

$$\frac{0.716-0.700}{0.716}=0.02 \qquad [6.11]$$

3. Further Comparisons of the Multi- and Single Point Methods

The analysis immediately above discloses that, when the BET C value is 100, the single point method using a relative pressure more than three times that required for monolayer coverage causes an error of only 2%. To further understand the BET equation and the relationship between the C value and the single point error it is useful to rewrite equation [5.38], the BET equation, as

$$\frac{W}{W_m}=\frac{C(P/P_0)}{[1+(C-1)P/P_0](1-P/P_0)} \qquad [6.12]$$

Using the method of partial fractions, the right side of equation [6.12] can be written as

$$\frac{C(P/P_0)}{[1+(C-1)P/P_0](1-P/P_0)}=\frac{X}{(1-P/P_0)}-\frac{Z}{1+(C-1)P/P_0} \qquad [6.13]$$

Using $X=Z=1$ as a solution gives

$$\frac{W}{W_m}=\frac{1}{1-P/P_0}-\frac{1}{1+(C-1)P/P_0} \qquad [6.14]$$

Equation [6.14] is the BET equation expressed as the difference between two rectangular hyperbole. If the value of C is taken as infinity then equation [6.14] immediately reduces to equation [6.3] the single point BET equation. The hyperbole referred to previously are shown in Figure 6-1.

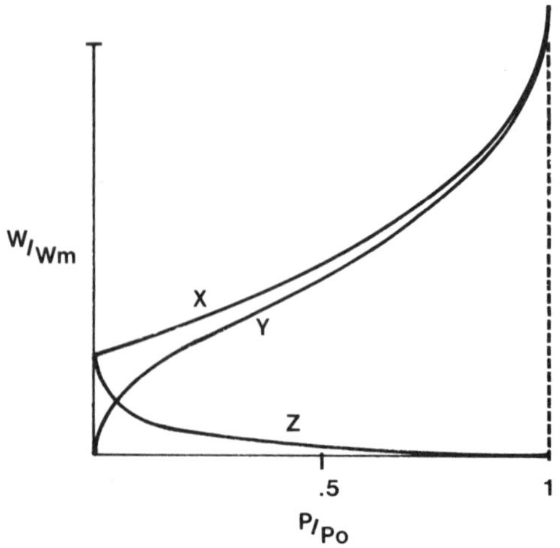

Figure 6-1 Plot of the hyperbole from equation [6.14] using an arbitrary C value.

$$\text{curve } X = \frac{1}{1 - P/P_0}; \text{ curve } Z = \frac{1}{1 + (C-1)P/P_0}; \text{ curve } Y = X - Z$$

As indicated by Figure 6-1, curve Y which is the BET curve for an arbitrary C value, approaches curve X, the single point curve, as the value of C increases. In the limiting case of C equal to ∞, the BET curve is coincident with the single point curve. For all other C values, the single point curve lies above the BET curve and their difference vanishes as the relative pressure approaches unity. Thus, as the value of C increases, the knee of the isotherm becomes sharper and moves toward lower relative pressures. For lower C values, curves X and Y diverge and higher relative pressures must be used to make single point surface areas conform to those obtained by the multipoint method.

The extent of divergence of curves X and Y is controlled, in a mathematical sense, by the second term in equation [6.13]. It is this term which contains the C value.

Table 6-1 indicates that, regardless of the C value, the relative error is reduced by using higher relative pressures. Similarly, Figure 6-1 shows that at sufficiently high relative pressures the BET curve and the single point curve merge regardless of the C value.

From this it would appear that all single point analyses should be performed at the highest possible relative pressures. Although theoretically sound, the use of relative pressures above 0.3 can lead to serious errors on the large number of samples that contain pores. In a later chapter the influence of pores is discussed but here it is sufficient to note that once condensation in pores commences, the BET equation, which deals only with adsorption, fails to adequately describe the state of the system. Ample evidence is available to indicate that many adsorbents possess pores that cause condensation at relative pressures as small as 0.3 and and in some cases at even lower values.[14] Therefore, relative pressures of 0.3 may be considered adequately high to give good agreement with multipoint measurements on most surfaces while avoiding condensation in all but microporous samples.

When used for quality control, the error associated with the single point method can be eliminated or greatly reduced if an initial multipoint analysis is performed to obtain the correct C value. Then equation [6.6] can be used to correct the results. Even an approximate value of C can be used to estimate the single point error. However, on the great majority of surfaces, the C value is sufficiently high to reduce the single point error to less than 5%.

CHAPTER SEVEN

Adsorbate Cross-Sectional Areas

Using the BET equation to determine W_m, the monolayer weight, and with reasonable estimates of thc adsorbate cross-sectional area, the sample surface area can be calculated by equation [5.13],

$$S_t = \frac{W_m \bar{N} \mathscr{A}}{\bar{M}} \times 10^{-20} \text{ m}^2 \qquad \text{[cf. 5.13]}$$

with W_m in grams, $\bar{M}$ as the adsorbate molecular weight, $\bar{N}$ is Avogadro's number, (6.02×10^{23} molecules per mole) and $\mathscr{A}$ in square angstroms per molecule, the sample surface area in square meters is given by equation [5.13]. Dividing by the sample weight converts S_t to S, the specific surface area.

1. Cross-Sectional Areas from the Liquid Molar Volume

A reasonable approximation to the cross-sectional area of the adsorbate was proposed by Emmett and Brunauer.[15] They assumed the adsorbate molecules to be spherical, and by using the bulk liquid properties they calculated the cross-sectional areas as

$$\mathscr{A} = \left(\frac{\bar{V}}{\bar{N}}\right)^{2/3} \times 10^{16} \text{ Å}^2 \qquad [7.1]$$

where $\overline{V}$ is the liquid molar volume. Equation [7.1] must be amended to reflect the molecular packing on the surface. Assuming that the liquid is structured as spheres with 12 nearest neighbors, 6 on a plane, in the usual close packed hexagonal arrangement and that the adsorbate has the same structure on the adsorbent surface, equation [7.1] becomes

$$\mathcal{A} = 1.09\left(\frac{\overline{V}}{\overline{N}}\right)^{2/3} \times 10^{16}\ \text{Å}^2 \qquad [7.2]$$

The factor of 1.09 is best explained by illustration. Figure 7-1 shows three rows of spherical adsorbate molecules adsorbed on the plane of the page in a close packed hexagonal structure. The length L is 1.09 times the width W. Therefore, the total area is 1.09 W^2. Thus, the factor 1.09 represents a packing factor unique for a close packed hexagonal arrangement.

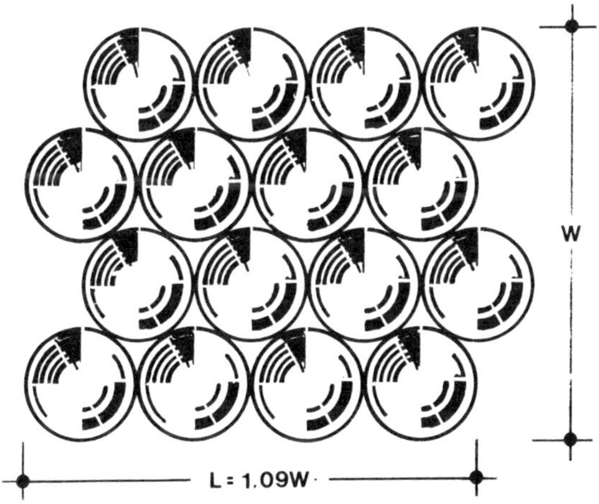

Figure 7-1 Sectional view of a close packed hexagonal arrangement of spheres. $L = 1.09W$.

That the adsorbate resides on the adsorbent surface with a structure similar to a plane of molecules within the bulk liquid, as depicted in Figure 7-1, is a simplified view of the real situation on the surfaces. Factors that make this model and therefore equation [7.2] of limited value include the following:

Any lateral motion of the adsorbate will be highly disruptive of any definite arrangement of adsorbate on the surface.

Complex molecules that rotate about several bond axis can undergo conformational changes on various surfaces and thereby also exhibit different cross-sectional areas.

Orientation of polar molecules produces different surface arrangements depending on the polarity of the adsorbent.

Strong interactions with the surface leads to localized adsorption which constrains the adsorbate to a specific site. The effective adsorbate cross-sectional area will then reflect the spacing between sites more than the actual adsorbate dimensions.

Fine pores may not be accessible to the adsorbate, thereby rendering inaccessible to measurement a substantial portion of the surface. This would be particularly true for large adsorbate molecules.

The conclusion to be reached from the aforesaid is that surface areas measured using cross-sectional areas calculated from equation [5.13] usually give different results depending upon the adsorbate used. If the cross-sectional areas are arbitrarily revised to give surface area conformity on one sample, the revised values generally will not give surface area agreement when the adsorbent is changed.

2. Nitrogen as the Standard Adsorbate

For surface area determinations the ideal adsorbate should exhibit BET C values sufficiently low to assure no localized adsorption. When the adsorbate is so strongly tied to the surface as to be constrained to a specific adsorption site the adsorbate cross-sectional area will be determined more by the adsorbent lattice structure than by the adsorbate dimensions. This type of "epitaxial" adsorption will lead to decreasing measured surface areas relative to the true BET value as the surface sites become more widely spaced.

On the other hand, if the C value is sufficiently small, the adsorbate lateral mobility on the surface will tend to disrupt any tendency for an organized structure to develop and the adsorbed layer might appear more as a two-dimensional gas.

In either of the preceding cases, very high or very low C values, any attempt to calculate the effective adsorbate cross-sectional areas from the bulk liquid properties will be subject to considerable error. Nitrogen, as an adsorbate, exhibits the unusual property that on almost all surfaces its C value is sufficiently small to prevent localized adsorption and yet adequately large to prevent the adsorbed layer from behaving as a two-dimensional gas.

Kiselev and Eltekov[16] established that the BET C value influences the adsorbate cross-sectional area. They measured the surface area of a number of adsorbents using nitrogen. When the surface areas of the same adsorbents were measured using *n*-pentane as the adsorbate, the cross-sectional areas of *n*-pentane had to be revised. It was found that the revised areas increased hyperbolically as the *n*-pentane C value decreased, as shown in Figure 7-2.

Figure 7-2 also shows that a plot of $(C^{1/2}-1)/(C-1)$ versus C gives a hyperbole which matches the data of Kiselev and Eltekov. Since $(C^{1/2}-1)/(C-1)$, $[(\theta_0)_m]$, can be calculated from a BET plot there exists a potential means of predicting the cross-sectional area variation relative to nitrogen. On surfaces that contain extensive porosity, which exclude larger adsorbate molecules from some pores while admitting smaller ones, it becomes even more difficult to predict any variation in the adsorbate cross-sectional area by comparison to a standard.[17,23]

In summary, it may be concluded that the uncertainty in calculating absolute adsorbate cross-sectional areas, the variation in cross-sectional areas with the BET C value, and the fact that on porous surfaces less area is available for larger adsorbate molecules all point to the need for a universal, although possibly arbitrary, standard adsorbate. The unique properties of nitrogen have led to its acceptance in this role with an assigned cross-sectional area of 16.2 Å^2 usually at its boiling point of $-195.6°$C.

3. Some Adsorbate Cross-Sectional Areas

Livingston[18] has assigned values of molecular cross-sectional areas to over twenty adsorbates based on what he calls best values.

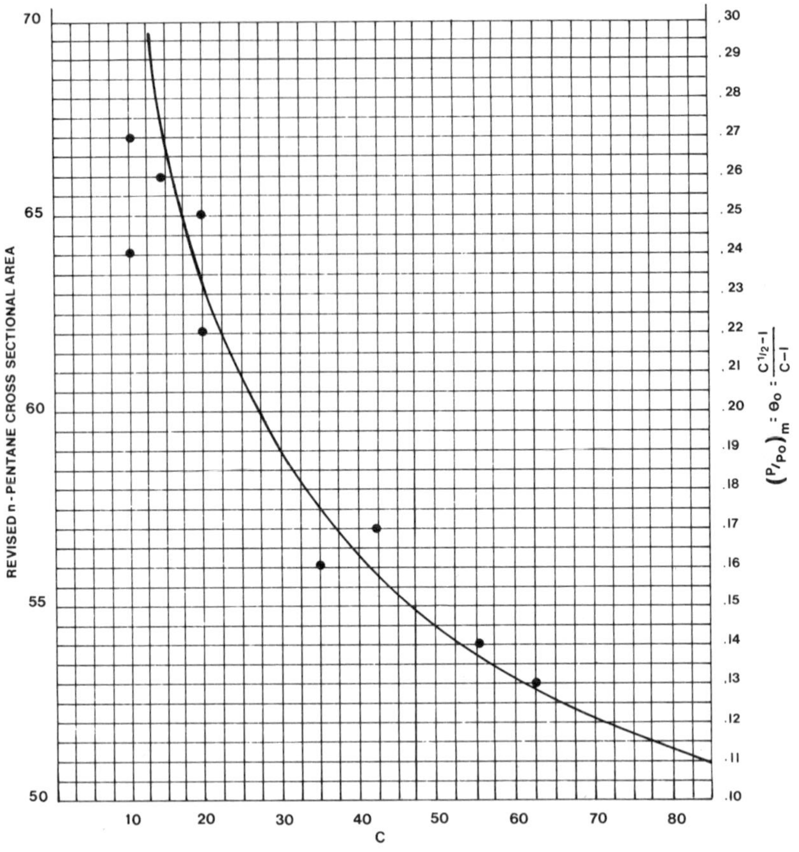

Figure 7-2 Variation of *n*-pentane cross-sectional area with the BET *C* value, (points).

These were obtained by assigning values to each adsorbate, as required, in order to make measured surface areas agree with the nitrogen value. For the purposes of comparison, Livingston assigned nitrogen the value of 15.4 Å^2 and placed the greatest weight on those adsorbents that exhibited no porosity.

Table 7-1 lists some approximate adsorbate cross-sectional areas. Because the adsorbates listed are used at various temperatures and on vastly different adsorbents the values listed are only approximate.

Table 7-1. Approximate Cross-Sectional Areas

Adsorbate	Area (Å²)	Reference
Argon	14.2	19
Ammonia	14.6	18
Benzene	40.0	20
Carbon dioxide	19.5	18
Carbon monoxide	16.3	18
Ethane	20.5	21
Krypton	19.5	22
n-Butane	46.9	23
n-Hexane	51.0	20
Nitrogen	16.2	24
Oxygen	14.1	25
Water	10.8	18
Xenon	2.5	26

With regard to cross-sectional areas, it must be kept in mind that the area occupied by a molecule or atom can often be many times its true area, and the terms "effective area" or "occupied area" are more appropriate and possibly less misleading.

CHAPTER EIGHT

Other Surface Area Methods

Because of its simplicity and straightforward applicability the BET theory is almost universally employed for surface area measurements. However, other methods and theoretical models have been developed which are briefly outlined in this chapter. No attempt is made to completely derive and discuss these alternate methods but only to present their essentials and indicate how they may be used to calculate surface areas.

1. Harkins and Jura Relative Method[27]

When a thin film of fatty acid is spread on the surface of water the surface tension of the water is reduced from γ_0 to γ. A barrier placed between pure water and water with a surface film will experience a pressure difference resulting from the tendency of the film to spread. This "surface pressure", π, is given by

$$\pi = \gamma_0 - \gamma \qquad [8.1]$$

Langmuir,[28] in 1917, constructed the "film balance" which permits experimental measurement of the "surface" or "spreading" pressure. Thus, it became possible to experimentally observe that adsorbed films pass through several states of molecular arrangement.[29] The various states resemble that of a two-dimensional gas, a low-density liquid, and finally a higher density or condensed liquid state. In the latter case the spreading pressure can be described by the linear relationship,

$$\pi = \alpha - \beta \mathscr{A} \qquad [8.2]$$

where $\mathscr{A}$ is the adsorbed molecules' effective cross-sectional area, and α and β are constants related to the film's compressibility.

A fundamental equation derived by Gibbs[30] is used to calculate the spreading pressure of films on solids where, unlike films on liquids, it can not be experimentally determined. Guggenheim and Adam[31] reduced Gibbs' general adsorption equation to equation [8.3] for the special case of gas adsorption.

$$d\pi = \frac{RTW}{\bar{M}S_t}\, d \ln P \qquad [8.3]$$

The terms in equation [8.3] have previously been defined as W= weight of adsorbate, $\bar{M}$ = adsorbate molecular weight, S_t = solid surface area, and P = equilibrium pressure.

Differentiating equation [8.2] and substituting for $d\pi$ in equation [8.3] yields

$$-\beta d\mathscr{A} = \frac{RTW}{\bar{M}S_t}\, d \ln P \qquad [8.4]$$

The implicit assumption made in deriving equation [8.4] is that an adsorbed gas on a solid surface behaves similar to a thin film of fatty acid on the surface of water.

Rewriting equation [5.13] as

$$\mathscr{A} = \frac{S_t \bar{M}}{W \bar{N}} \qquad [\text{cf. } 5.13]$$

and differentiating $\mathscr{A}$ with respect to W yields

$$d\mathscr{A} = -\left(\frac{S_t \bar{M}}{\bar{N}}\right) \frac{dW}{W^2} \qquad [8.5]$$

Replacing $d\mathscr{A}$ in equation [8.4] by equation [8.5] and rearranging terms gives

$$d \ln P = \left(\frac{\beta \overline{M}^2 S_t^2}{RT\overline{N}}\right) \frac{dW}{W^3} \qquad [8.6]$$

Integrating equation [8.6] and subtracting $\ln P_0$ from each side yields

$$\frac{\ln P}{P_0} = -\left(\frac{\beta \overline{M}^2 S_t^2}{2RT\overline{N}}\right) \frac{1}{W^2} + \text{Const.} - \ln P_0 \qquad [8.7]$$

or

$$\frac{\ln P}{P_0} = A - \frac{B}{W^2} \qquad [8.7a]$$

According to equation [8.7], the Harkins Jura equation, a plot of $\ln P/P_0$ versus $1/W^2$ should give a straight line with the slope equal to $-B$ and the intercept equal to A. The surface area is then calculated as

$$S_t = \frac{1}{\overline{M}} \left(\frac{2\overline{N}RT}{\beta}\right)^{1/2} B^{1/2} \qquad [8.8]$$

or

$$S_t = KB^{1/2} \qquad [8.8a]$$

where

$$K = \frac{1}{\overline{M}} \left(\frac{2\overline{N}RT}{\beta}\right)^{1/2} \qquad [8.9]$$

The term K in equation [8.9] is the Harkins–Jura (HJ) constant and is assumed to be independent of the adsorbent and dependent only on the adsorbate.

In some instances Harkins and Jura found two or more linear regions of different slopes when $\ln P/P_0$ is plotted versus W^{-2}. This indicates the existence of two or more liquid condensed states where different molecular packing occurs. When this situation appears, the slope that gives best agreement with an alternate method, for example, the BET method, must be chosen; alternatively the temperature or the adsorbate can be changed to eliminate the ambiguity.

Emmett[32] has shown that linear HJ plots are obtained for relative pressures between 0.05 and about 0.30, the usual BET range, when the BET C value lies between 50 and 250. For C values below 10 the linear HJ region lies above relative pressures of 0.4, and for $C = 1000$ the HJ plot is linear between 0.01 and 0.18. Further, Emmett has found that the nitrogen cross-sectional area had to be adjusted from 13.6 Å^2 to 18.6 Å^2 as C varied between 50 and 250 in order to give the same surface area as the HJ method using 4.06 as the value for K.

2. Harkins and Jura Absolute Method[33]

In addition to their relative method, Harkins and Jura have also developed an absolute method for surface area measurement which is independent of the adsorption isotherm and is based entirely upon calorimetric data. Consider a system in which 1 gram of solid and α moles of vapor are transformed into a state in which the solid is immersed into a liquid made by condensing the vapor. As shown in Figure 8-1, this transformation can be accomplished along two possible paths.

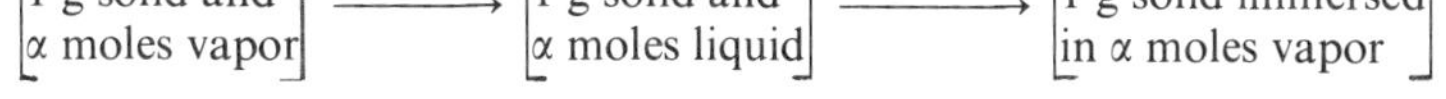

Figure 8-1

The terms ΔH_i, L, ΔH_{sv}, and i used in Figure 8-1 are all enthalpy changes defined as follows: ΔH_i is the heat of immersion of the solid into the liquid, L is the latent heat of condensation, ΔH_{sv} is the heat of adsorption when the solid is equilibrated with saturated vapor, and i is the heat liberated when solid in equilibrium with saturated vapor is immersed into liquid. Using Hess's law of heat summation

$$\Delta H_1 + \Delta H_2 = \Delta H_3 + \Delta H_4 + \Delta H_5 \qquad [8.10]$$

then

$$\Delta H_i = (\Delta H_{sv} - \alpha_1 L) + i \qquad [8.11]$$

where the quantity $(\Delta H_{sv} - \alpha_1 L)$ is the integral heat of adsorption. This value as well as the value of i can be measured using calorimetry. The value of i is actually zero if the isotherm approaches the ordinate asymptotically. If the isotherm cuts the ordinate at a finite angle, i will be finite but small.

For powder samples that differ only in their surface area the heat of immersion will be proportional to the surface area. Thus,

$$H_i = h_i S \qquad [8.12]$$

where h_i is the heat of immersion per unit area of solid.

Gregg and Sing[34] have tabularized values of h_i for various solids and liquids. However, values of h_i found in the literature must be used with great caution since an error of even 1 calorie would produce a surface area error of 42 m^2 with a typical h_i value of 100 erg/cm^2. Serious errors can also be encountered if close attention is not paid to the liquid purity since the measured heat may reflect a quantity of strongly adsorbed impurities. With regard to the solid itself, even identical chemical analysis will not guarantee that two samples will have the same h_i value. The same material prepared differently or with different histories can possess varying amounts of lattice strain which will produce meaningful differences in h_i values. Samples with differing pore sizes but otherwise identical will also exhibit varying h_i values due to the variation in the potential fields within the pores. Ideally therefore, it is desirable to work with carefully annealed nonporous materials when using the heat of immersion method. These types of samples are rarely encountered and their low surface areas will yield small heat values, making the experimental determination of their surface areas a tedious and difficult procedure. These difficulties indicate clearly that the immersion method is of questionable value as a rapid or routine method for surface area measurements.

An important contribution made by the Harkins and Jura absolute method must not be overlooked. Their measurements of some specific surface areas give confirmation to the value of 16.2 $Å^2$ for the cross-sectional area of nitrogen. This value, when employed with the BET theory, gave exactly the same specific surface area as the HJ absolute method.

3. Permeametry

Consider a gas near ambient pressure and temperature forced, by a small pressure gradient, to flow through the channels of a packed bed of powder. At room temperature, a gas molecule can be adsorbed on a solid surface for an extremely short time but not less than the time required for one vibrational cycle or about 10^{-13} seconds. When the adsorbed molecule leaves the surface it will, on the average, have a zero velocity component in the direction of flow. After undergoing one or several gas phase collisions it will soon acquire a drift velocity equal to the linear flow velocity. These collisions and corresponding momentum exchanges will occur within one, or at most a few, mean free path lengths from the surface with the net effect of decreasing the linear velocity of those molecules flowing near the surface. As the area to volume ratio of the channels in the powder bed increases, the viscous drag effect will also increase and the rate of flow, for a given pressure gradient, will decrease. Clearly then, the surface area of the particles which constitute the channel walls is related to the gas flow rate and pressure gradient. This type of flow is called viscous flow and is described by Poiseuille's[35] law.

When the mean free path is approximately the same as the channel diameter, as is the case with coarse powders and reduced gas pressure or fine powders and the gas at atmospheric pressures, the gas will behave as though there were slippage at the channel walls. This occurs because collisions between molecules rebounding from a wall and the flowing molecules occur uniformly across the diameter of the channel. Therefore, there appears to be no preferential retardation of flow near the channel wall when compared to the center of the channel.

A third type of flow occurs at significantly reduced pressures where the mean free path of the gas molecules is greater than the channel diameter. Viscosity plays no part in this type of flow since the molecular collisions with the channel walls far outnumber the gas phase collisions. This type of molecular flow is a diffusion process.

Depending upon which of the preceding three types of flow is employed, somewhat different measurement of a sample's surface area is obtained. Viscous flow measurements tends to ignore blind pores and measures only the "envelope" area of the particles. At the other extreme, diffusional flow senses the blind pores and often gives good agreement with surface areas measured by the BET method. At intermediate pressures, with so called slip flow, the very small blind pores are ignored while the larger ones can contribute to the measured surface area.

Darcy's law[36] asserts that the average flow velocity of a fluid through a packed bed is proportional to the pressure gradient across the bed and is inversely proportional to the length of the bed. Thus,

$$v = K\frac{\Delta p}{l} \tag{8.13}$$

J. L. M. Poiseuille[35] has shown that the viscosity η of a fluid can be expressed in terms of the volume V (cm^3) flowing through a tube of radius r (cm) and length l (cm) in time t (sec) under a pressure gradient of ΔP dyne cm^{-2}.

$$\eta = \frac{\pi r^4 t \Delta p}{8Vl} \tag{8.14}$$

Equation [8.14] can be rearranged, by substituting $\pi r^2 l$ for V and v for l/t.

$$v = \frac{D^2 \Delta p}{32\eta l} \tag{8.15}$$

where v is the linear flow velocity and D is the tube diameter. The similarity in form of equations [8.13] and [8.15] suggests that flow through a packed powder bed is equivalent to fluid flow through many small capillaries or channels. Kozeny[37] recognized this equivalency when he derived an equation for flow through a packed bed. Kozeny used the ratio of volume to the area of a tube, viz.,

$$\frac{V}{A}=\frac{D}{4} \qquad [8.16]$$

If many parallel capillaries of equal length and diameters are used to carry the fluid, the ratio of V to A remains constant and equation [8.16] becomes

$$D_e=\frac{4V}{A} \qquad [8.17]$$

where D_e is the diameter equivalent to one large tube.

In a powder bed the void volume V_v is defined as the volume not occupied by solids. Further, the porosity p is defined by

$$p=\frac{V_v}{V_s+V_v} \qquad [8.18]$$

where V_s+V_v is the total volume of the powder bed and V_s is the volume occupied by the solid. Rearranging equation [8.18] gives

$$V_v=\left(\frac{p}{1-p}\right)V_s \qquad [8.19]$$

Substituting V_v from equation [8.19] into equation [8.17] gives

$$D_e=4\left(\frac{p}{1-p}\right)\frac{V_s}{A} \qquad [8.20]$$

and replacing D in equation [8.15] with D_e preceding, yields

$$v_p=\left(\frac{V_s^2\Delta P}{A^2 2\eta l_p}\right)\left(\frac{p}{1-p}\right)^2 \qquad [8.21]$$

The term v_p is the average flow velocity through the channels in the bed and l_p is the average channel length.

Because the approach velocity v, that is, the gas velocity prior to entering the powder bed, is experimentally simple to measure, it is desirable to state equation [8.21] in terms of v rather than v_p.

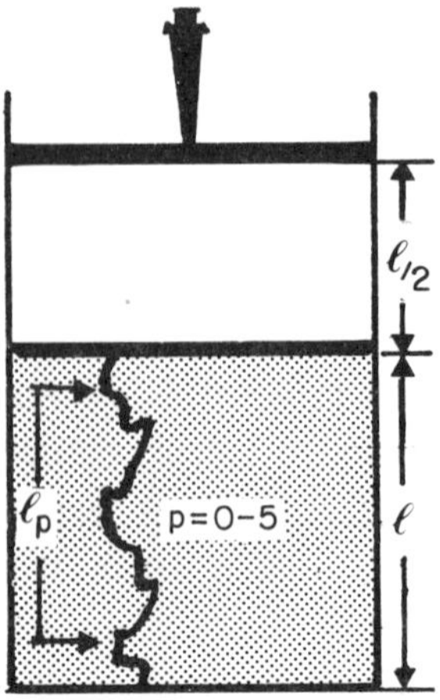

Figure 8-2 Piston displacement forcing gas to flow along a tortuous path through a packed powder bed.

In order to make this conversion consider Figure 8-2. The unshaded area represents gas contained between a piston and the surface of the powder bed. If the porosity of the bed is 0.5, then in order to contain an equal volume of gas, the powder bed must be twice as long as the distance the piston can move. Clearly then, if the piston is displaced downward at a uniform rate the linear velocity through the bed must be twice the approach velocity, or

$$v_p = \frac{v}{p} \tag{8.22}$$

However, equation [8.22] assumes that the channels in the powder bed are straight through. In fact they are tortuous in shape as indicated in the diagram. Actually, in the time required to travel through the powder bed an element of gas volume traverses a distance l_p/l greater than the length of the powder bed. This increases v_p and equation [8.21] is corrected to

$$v_p = \frac{v}{p}\left(\frac{l_p}{l}\right) \tag{8.23}$$

Replacing v_p in equation [8.21] with v from equation [8.23] gives

$$v = p\left(\frac{l}{l_p}\right)\left(\frac{V_s^2 \Delta P}{A^2 2\eta l_p}\right)\left[\frac{p^2}{(1-p)^2}\right] \qquad [8.24]$$

and

$$A^2 = \frac{1}{2}\left(\frac{l}{l_p}\right)^2\left(\frac{V_s^2 \Delta P}{\eta v l}\right)\left[\frac{p^3}{(1-p)^2}\right] \qquad [8.25]$$

Further, if V_s^2 is replaced by the ratio of mass to density, equation [8.25] becomes

$$\left(\frac{A}{m}\right)^2 = \frac{1}{2}\left(\frac{l}{l_p}\right)^2\left(\frac{\Delta P}{\rho^2 \eta l v}\right)\left[\frac{p^3}{(1-p)^2}\right] \qquad [8.26]$$

The area A represents the total surface area of all the channels in the powder bed which offer resistance to flow. This excludes the surface area generated by close-ended or blind pores and crevices since they make no contribution to the frictional area. It follows then that equation [8.26] will give surface area values always less than BET or other adsorption techniques.

Recognizing that the ratio A/m in [8.26] is the specific surface area and calling $2(lp/l)^2$ the "aspect factor" f, which is determined by the particle's geometry, equation [8.26] can be written as

$$S^2 = \left(\frac{\Delta P}{l}\right)\left(\frac{p^3}{(1-p)^2}\right)\left(\frac{1}{f \rho^2 \eta v}\right) \qquad [8.27]$$

Equation [8.27] will hold for incompressible fluids and for compressible fluids with small values of ΔP. If the pressure gradient across the bed is large and the fluid compressible equation [8.27] takes the form

$$S^2 = \left(\frac{P_2^2 - P_1^2}{2P_1}\right)\left(\frac{p^3}{(1-p)^2}\right)\left(\frac{1}{f \rho^2 l \eta v}\right) \qquad [8.28]$$

where P_1 and P_2 are the exit and inlet pressures, respectively.

Experimentally the terms in equations [8.27] or [8.28] can be determined as follows:

1. ΔP can be measured by placing a sensitive pressure gauge or manometer at the inlet to the powder bed while venting to atmospheric pressure.
2. ρ, the true powder density, can be determined by a variety of methods, one of which is discussed in Chapter 19.
3. v can be determined by dividing the volumetric flow rate by the powder bed's cross-sectional area.
4. p is calculated by subtracting the volume of powder from the total bed volume and dividing by the total bed volume.

The aspect factor, f, equal to $2(l_p/l)^2$ arises from the size and shape of the cross-sectional areas that make up the channels. Therefore, f is highly dependent upon the particle size and shape. Carman[38] studied numerous materials and found a value of 5 suitable in most cases. Various theoretical and experimental values for f usually are found to be in the range of 3 to 6.

Permeation measurements will be prone to serious errors if the average channel diameter is not within a factor of two of the largest diameter. Large-diameter channels tend to give area to volume ratios excessively low (cf. equation 8.17) while their contribution to the average flow rate is in excess of their number. Also, agglomerated particles behave as one particle, the flow measuring only the envelope of the aggregate.

Several modifications of Poiseuille's equation have been attempted by various authors[39-41] to describe permeability in the transitional region between viscous and diffusional flow. The assumptions underlying these modifications are often questionable and the results obtained offer little or no theoretical or experimental advantage over the BET theory for surface area measurements. Allen[42] discusses these modifications as well as diffusional flow at low pressures.

Viscous flow permeametry measured near atmospheric pressure offers the advantages of experimental simplicity and a means of measuring the external or envelope area of a powder sample which is otherwise not readily available by any adsorption method. The usefulness of measuring the external surface area rather than the BET or total surface area becomes evident if the data is to be correlated with fluid flow through a powder bed or with the average particle size.

CHAPTER NINE

Pore Analysis by Adsorption

1. The Kelvin Equation

Adsorption studies leading to measurements of pore size and pore size distributions generally make use of the Kelvin[43] equation which relates the equilibrium vapor pressure of a curved surface, such as that of a liquid in a capillary or pore, to the equilibrium pressure of the same liquid on a plane surface. Equation [9.1] is a convenient form of the Kelvin èquation:

$$\ln \frac{P}{P_0} = -\frac{2\gamma \bar{V}}{rRT} \cos \theta \qquad [9.1]$$

where P is the equilibrium vapor pressure of the liquid contained in a narrow pore of radius r and P_0 is the equilibrium pressure of the same liquid exhibiting a plane surface. The terms γ and $\bar{V}$ are the liquid's surface tension and molar volume, respectively, and θ is the contact angle at which the liquid meets the pore wall.

In a pore, the overlapping potentials of the walls more readily overcome the translational energy of an adsorbate molecule so that condensation will occur at a lower pressure than is normally required on an open or plane surface. Thus, as the relative pressure is increased condensation will occur first in pores of smaller radii and will progress into the larger pores until, at a relative pressure of unity, condensation will occur on those surfaces where the radius of curvature is essentially infinite. Conversely, as the relative pressure decreases, evaporation will occur progressively out of pores with decreasing radii.

In order to derive the Kelvin equation upon thermodynamic grounds, consider the transfer of dn moles of vapor in equilibrium with the bulk liquid at pressure P_0 into a pore where the equilibrium pressure is P. This process consists of three steps: evaporation from the bulk liquid, expansion of the vapor from P_0 to P, and condensation into the pore. The first and third of these steps are equilibrium processes and are therefore accompanied by a zero free energy change, whereas the free energy change for the second step is described by

$$dG = \left(RT \ln \frac{P}{P_0} \right) dn \qquad [9.2]$$

When the adsorbate condenses in the pore it does so on a previously adsorbed film thereby decreasing the film–vapor interfacial area. The free energy change associated with the filling of the pore is given by

$$dG = -(\gamma \cos \theta)\, dS \qquad [9.3]$$

where γ is the surface tension of the adsorbed film, assumed to be identical with that of the liquid, dS is the change in interfacial area, and θ is the wetting angle which is taken to be zero since the liquid is assumed to completely wet the adsorbed film. Figure 9-1 illustrates the difference between a zero and a finite wetting angle.

Equations [9.2] and [9.3], when combined using the assumption of a zero wetting angle, yield

$$\frac{dn}{dS} = \frac{-\gamma}{RT \ln P/P_0} \qquad [9.4]$$

The volume of liquid adsorbate that condenses in a pore of volume V_p is given by

$$dV_p = \overline{V} dn \qquad [9.5]$$

where $\overline{V}$ is the liquid adsorbate molar volume. Substituting equa-

Figure 9-1 Wetting angles.

tion [9.5] into equation [9.4] gives

$$\frac{dV_p}{dS}=\frac{-\gamma\bar{V}}{RT\ln P/P_0} \quad [9.6]$$

The ratio of volume to area within a pore depends upon the pore geometry. For example, the volume to area ratio for a cylinder, parallel plates, and a sphere are, respectively, $r/2$, $r/2$, and $r/3$, where r is the cylinder and sphere radius or the distance of separation between parallel plates. If the pore shapes are highly irregular or consist of a mixture of regular geometries, the volume to area ratio can be too complex to express mathematically. In these cases, or in the absence of specific knowledge of the pore geometry, the assumption of cylindrical pores is usually made, in which case equation [9.6] becomes

$$\ln\frac{P}{P_0}=\frac{-2\gamma\bar{V}}{rRT} \quad [9.7]$$

Equation [9.7] is the working equation for pore size analysis by adsorption unless more specific information is available regarding pore geometry and the wetting angle.

Of the five isotherm classifications depicted in Figure 4-1 the types I, IV, and V isotherms are associated with porosity. The type I isotherm usually corresponds to microporosity, that is, pores of diameters only slightly larger than adsorbate molecules. The types

IV and V isotherms are associated with pores ranging from about fifteen to several hundred angstroms with the type IV isotherm far more frequently encountered.

The types IV and V isotherms resemble the type II and type III isotherms, respectively, at the bottom or low-pressure end. However, at the high-pressure end they turn toward the $P/P_0=1$ line and as indicated in Figure 9-2 the isotherm can approach the saturated pressure axis asymptotically (D–G) or at a finite angle (D–E).

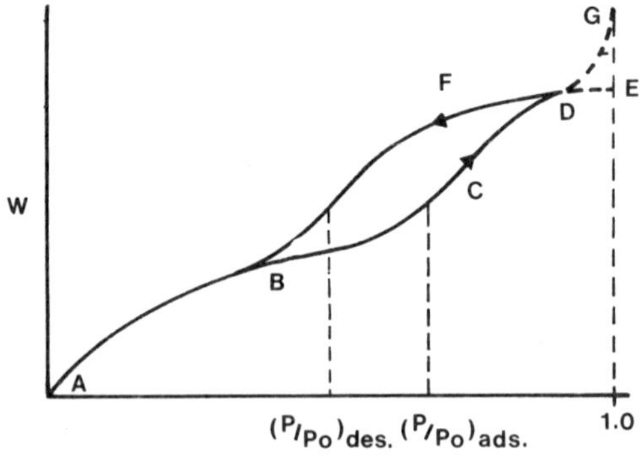

Figure 9-2 Typical type IV adsorption and desorption isotherms showing hysteresis.

In those cases where the isotherm cuts the saturated pressure line at a finite angle the liquid condensed in the pores does not completely wet the previously adsorbed film. This condition is presumed to arise because the film persists in exhibiting a different structure than the liquid because of the extended influence of the adsorbent's surface potential. In this case it would be more correct to use equation [9.1] which includes the wetting angle rather than equation [9.7] which assumes a zero wetting angle. However, for wetting angles even as large as 20° the value of $\cos\theta$ is 0.94. Thus, the error caused by assuming a zero wetting angle will generally be small, particularly when compared to the errors introduced because of lack of knowledge regarding the exact pore shape.

When the isotherm approaches the saturated pressure line

asymptotically, which is the prevalent condition, wetting of the previously adsorbed film by the adsorbate condensed in the pores is assured and the wetting angle may be taken with certainty as zero.

2. Adsorption Hysteresis

Figure 9-2 also shows hysteresis which is typical of type IV isotherms. The line *BCD* indicates the path traversed along the adsorption isotherm as the relative pressure is increased, whereas the line *DFB* shows the path followed along the desorption isotherm as the relative pressure is reduced. The presence of the hysteresis loop introduces a considerable complication, in that within the region of the hysteresis loop there are two relative pressure values corresponding to a given quantity adsorbed with the lower value always residing on the desorption isotherm. As shown in Figure 9-2, the quantity W is adsorbed at a lower relative pressure on the desorption isotherm than on the adsorption curve. The pore radius corresponding to the quantity adsorbed must be single-valued and some criteria have to be established as to which value of relative pressure is to be employed in the Kelvin equation. As shown previously (cf. equation [9.2]) the molar free energy change accompanying the condensation of vapor into a pore during adsorption is given by

$$\Delta G_{\text{ads}} = RT\,(\ln P_{\text{ads}} - \ln P_0) \qquad [9.8]$$

For the same quantity on the desorption isotherm the corresponding free energy change is

$$\Delta G_{\text{des}} = RT(\ln P_{\text{des}} - \ln P_0) \qquad [9.9]$$

Since $P_{\text{des}} < P_{\text{ads}}$ it follows that $\Delta G_{\text{des}} < \Delta G_{\text{ads}}$. Therefore, the desorption value of relative pressure corresponds to the more stable adsorbate condition and the desorption isotherm should, with certain exceptions, be used for pore size analysis.

Several theories have been formulated to explain the difference between the state of the adsorbate during adsorption and desorption. Zsigmondy[44] postulated that hysteresis was caused by a difference of contact angle during adsorption and desorption.

McBain[45] accounted for hysteresis by assuming the pores contained a narrow opening and a wide body, the so-called "bottle-neck" shape. His model asserts that during adsorption the wide inner portion of the pore is filled at high relative pressures but cannot empty until the narrow neck of the pore empties at lower relative pressures during desorption. It follows, therefore, that for "bottleneck" pores the adsorption isotherm corresponds to the equilibrium condition.

Cohan[46] assumed that condensation occurs by filling the pore from the wall inward which, for a cylindrical pore, would give a cylindrical shaped meniscus whereas evaporation occurs from a hemispherical meniscus once the pore is filled.

Foster[47] explained hysteresis by considering the pores to fill by adsorption on the walls while emptying by evaporation from a spherical meniscus.

These and other theories[48] may each describe a unique condition leading to hysteresis and indeed there may be no single mechanism that can universally explain hysteresis.

3. Types of Hysteresis

Rarely have hysteresis loops been found which have not closed, at the low end, by relative pressures of 0.3. According to the Kelvin equation, pore radii corresponding to relative pressures less than 0.3 would be smaller than 15 Å. Since monolayer formation is usually complete when the relative pressure reaches 0.3, the radius available for condensation would be diminished by the thickness of the monolayer or by about two molecular diameters. The radius of this center core would then be approximately one or two molecular diameters. It would be difficult to establish the validity of the Kelvin equation for pores this small, and it is entirely possible that adsorption and desorption in micropores proceed by the same path thereby precluding hysteresis.

At relative pressures above 0.3, de Boer[49] has identified five

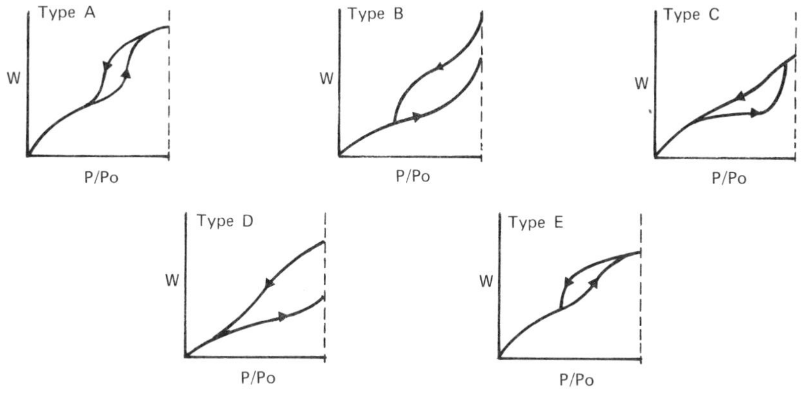

Figure 9-3 De Boer's five types of hysteresis.

types of hysteresis loops which he has correlated with various pore shapes. Figure 9-3 shows idealizations of the five types of hysteresis.

Type A hysteresis is due principally to cylindrical pores open at both ends. According to Cohan,[46] type A hysteresis is caused by condensation producing a cylindrical meniscus with one radius of curvature equal to the pore radius, less the thickness of previously condensed film, and the other radius is the length of the pore or essentially infinite. During desorption, Cohan views the meniscus as being hemispherical. According to this model, as the cylindrical film increases in depth during adsorption, the changes in the remaining volume and area are given, respectively, by

$$dV = -2\pi r_k l dr_k \qquad [9.10]$$

and

$$dS = -2\pi l dr_k \qquad [9.11]$$

where r_k is the radius of the unfilled part, or center core, of the cylindrical pore and l is the pore length. Combining equations [9.10] and [9.11] gives

$$\frac{dV}{dS} = r_k \qquad [9.12]$$

Placing the above value for dV/dS into equation [9.6] yields a modified form of the Kelvin equation, namely,

$$\ln\left(\frac{P}{P_0}\right)_{\text{ads}} = -\frac{\gamma\bar{V}}{r_k RT} \qquad [9.13]$$

During desorption the center core evaporates from a hemispherical meniscus; therefore, the Kelvin equation is applicable

$$\ln\left(\frac{P}{P_0}\right)_{\text{des}} = -\frac{2\gamma\bar{V}}{r_k RT} \qquad [\text{cf. } 9.7]$$

From equations [9.13] and [9.7], it is evident that condensation into and evaporation out of the inner core will occur when

$$\left(\frac{P}{P_0}\right)_{\text{ads}}^2 = \left(\frac{P}{P_0}\right)_{\text{des}} \qquad [9.14]$$

Therefore, for a real system of pores of approximately cylindrical geometry it is expected that pores of a given radius will fill at a higher relative pressure than they will empty.

The type B hysteresis curve is associated with slit-shaped pores or the space between parallel plates.

Type C hysteresis is produced by a mixture of tapered or wedge-shaped pores with open ends.

Type D curves are also produced by tapered or wedge-shaped pores but with narrow necks at one or both open ends.

Type E hysteresis results from McBain's "bottleneck" pores. In pores of this shape, emptying of the wide portion will be delayed during desorption until the narrow neck can evaporate. Therefore, the desorption curve exhibits a small slope at high relative pressures and a large slope where the wide part of the pore evaporates.

4. Total Pore Volume

If W_s is the adsorbate weight at saturation (i.e., when $P/P_0 = 1$), then $W_s/\rho_l = V_l$, where V_l and ρ_l are the volume of liquid adsorbate at saturation and the liquid density, respectively. Various studies[50–53] have shown that at saturation, the liquid volume of different adsorbates, when measured on porous adsorbents, is essentially constant and is independent of the adsorbate. This constancy of adsorbed liquid at saturation is known as the Gurvitsh[54] rule and provides direct evidence that the pores are filled with liquid adsorbate at saturated vapor pressures.

To calculate the total pore volume it is necessary to measure the total adsorbate volume at a relative pressure above the point where the hysteresis loop closes. In general, a relative pressure as close to unity as possible should be chosen in order to include the large radii pores in the measurement. For example, if W_a grams of nitrogen is adsorbed at $P/P_0 = 0.99$ then the corresponding volume of pores V_p and the largest pore radii r are given by

$$V_p = \frac{W_a}{\rho_l} \qquad [9.15]$$

and from the Kelvin equation

$$r = \frac{-2(8.85)(34.6)}{(8.314 \times 10^7)(77)(2.303)\log 0.99} = 950 \times 10^{-8} \text{ cm} \qquad [9.16]$$

where 8.85 erg cm^{-2} is the surface tension and 34.6 cm^3 is the molar volume of liquid nitrogen at 77°K. Equations [9.15] and [9.16] state that the total volume of all the pores up to 950 Å is V_p cm^3. An implicit assumption, in all the foregoing, is that no surface other than the inner walls of the pores exists. Using this assumption and that of cylindrical geometry, the average pore radius r_p can be calculated from the ratio of the total pore volume and BET surface area from the following equation:

$$\frac{V_p}{S_{BET}} = \frac{r_p}{2} \qquad [9.17]$$

5. Pore Size Distributions

On thermodynamic grounds pore size distributions are measured using the desorption isotherm (see equations [9.8] and [9.9]). The exception to this are "bottleneck" pores exhibiting type E hysteresis. In this case the equilibrium isotherm is that of adsorption because the unstable state is associated with the wide portion of the pore being unable to evaporate until the narrow neck empties. Regardless, however, of which isotherm is used the mathematical treatment remains the same.

For nitrogen as the adsorbate at its normal boiling point of 77° K, the Kelvin equation can be written as

$$r_k = \frac{4.15}{\log P_0/P} \ (\text{Å}) \tag{9.18}$$

using the physical properties given in equation [9.16]. The term r_k indicates the radius into which condensation occurs at the required relative pressure. This radius, called the Kelvin radius or the critical radius, is not the actual pore radius because some adsorption has occurred on the pore wall, prior to condensation, leaving a center core of radius r_k. Alternatively, during desorption, an adsorbed film remains on the pore wall when evaporation of the center core takes place.

If the depth of the adsorbed film when condensation or evaporation occurs is t, then the actual pore radius r_p is given by

$$r_p = r_k + t \tag{9.19}$$

Equation [9.18] can be used to calculate r_k but some means of evaluating t is required if the pore radius is to be determined.

Using the assumption that the adsorbed film depth in a pore is the same as on a plane surface for any value of relative pressure, one can write

$$t = \left(\frac{W_a}{W_m}\right)\tau \tag{9.20}$$

where W_a and W_m are, respectively, the quantity adsorbed at a particular relative pressure and the weight corresponding to the BET monolayer. Essentially, equation [9.20] asserts that the thickness of the adsorbed film is simply the number of layers times the thickness τ of one layer regardless of whether the film is in a pore or on a plane surface.

The value of τ can be calculated by considering the area and volume occupied by one mole of liquid nitrogen if it were spread over a surface to the depth of one molecular layer.

$$S=(16.2)(6.02\times 10^{23})=97.5\times 10^{23}\ \text{Å}^2 \qquad [9.21]$$
$$\bar{V}=(34.6\times 10^{24})\ \text{Å}^3$$

then

$$\tau=\frac{\bar{V}}{S}=3.54\ \text{Å} \qquad [9.23]$$

this value of 3.54 Å is somewhat less than the diameter of a nitrogen molecule based upon the cross-sectional area of 16.2 Å^2. This is as it should be because the liquid structure is considered to be close packed hexagonal and each nitrogen molecule sits in the depression between three molecules in the layers above and below.

Equation [9.20] can now be written as

$$t=\left(\frac{W_a}{W_m}\right)3.54\ \text{Å} \qquad [9.24]$$

On nonporous surfaces it has been shown that when W_a/W_m is plotted versus P/P_0 the data all approximately fit a common type II curve above a relative pressure of 0.3.[55-60] This means that when $W_a/W_m=3$, for example, the adsorbed layer thickness t will be 10.62 Å regardless of the adsorbent. The common curve is described closely by the Halsey[61] equation which for nitrogen can be written as

$$t=3.54\left[\frac{5}{2.303\log P_0/P}\right]^{1/3} \qquad [9.25]$$

To calculate the pore size distribution consider the work sheet shown as Table 9-1, and the corresponding explanation of each column. The adsorbed volumes are from a hypothetical isotherm. The procedure used is the numerical integration method of C. Pierce[58] as modified by Orr and Dalla Valle[62] with regard to calculating the depth of the adsorbed film. This method as well as the Barrett, Joyner, and Halenda (BJH)[63] numerical integration method takes advantage of Wheeler's[64] theory that condensation occurs in the pores when a critical relative pressure is reached which corresponds to the Kelvin radius r_k. This model also assumes that a

Table 9-1. Pore Size Distribution Work Table

1	2	3	4	5	6	7	8	9	10	11	12	13	14
$\frac{P}{P_0}$	V_{gas} STP ($cm^3\ g^{-1}$)	r_k (Å)	t (Å)	r_p (Å)	$\bar{r}_k$ (Å)	$\bar{r}_p$ (Å)	Δt (Å)	ΔV_{gas} STP ($cm^3\ g^{-1}$)	ΔV_{liq} $\times 10^3$ ($cm^3\ g^{-1}$)	$\Delta t \Sigma S$ $\times 10^3$ ($cm^3\ g^{-1}$)	V_p $\times 10^3$ ($cm^3\ g^{-1}$)[a]	S (m^2)	ΣS (m^2)[b]
.99	161.7	950	28.0	978									
					711	737	5.8	0.2	0.31	0	0.33	.01	.01
.98	161.5	473	22.2	495									
					394	414	2.8	0.5	0.77	0.00	0.85	.04	.05
.97	161.0	314	19.4	333									
					250	268	3.2	0.8	1.23	0.02	1.39	.10	.15
.95	160.2	186	16.2	202									
					138	153	3.4	1.4	2.16	0.05	2.59	.34	.49
.90	158.8	90.7	12.8	104									
					74.8	87.0	1.7	1.6	2.46	0.08	3.22	.74	1.23
.85	157.2	58.8	11.1	69.9									
					50.8	61.4	1.1	2.0	3.08	0.14	4.29	1.40	2.63
.80	155.2	42.8	10.0	52.8									
					39.7	49.5	.5	2.3	3.54	0.13	5.30	2.14	4.77
.77	152.9	36.6	9.5	46.1									
					34.9	44.3	.3	4.0	6.16	0.14	9.70	4.38	9.15
.75	148.9	33.2	9.2	42.4									
					31.8	40.9	.3	3.8	5.85	0.27	9.23	4.51	13.66
.73	145.1	30.4	8.9	39.3									
					29.2	38.0	.2	4.2	6.47	0.27	10.50	5.53	19.19
.71	140.9	27.9	8.7	36.6									
					26.9	35.4	.3	5.0	7.70	0.58	12.33	6.97	26.16
.69	135.9	25.8	8.4	34.2									
					24.9	33.2	.2	5.9	9.09	0.52	15.24	9.18	35.34
.67	130.0	23.9	8.2	32.1									

					23.1	31.2	.2	6.1	9.39	0.71	15.83	10.15	45.49
.65	123.9	22.2	8.0	30.2									
					21.5	29.4	.2	6.6	10.16	0.91	17.29	11.76	57.25
.63	117.3	20.7	7.8	28.5									
					20.0	28.8	.1	7.2	11.09	0.57	21.81	15.15	72.40
.61	110.1	19.3	7.7	27.0									
					18.7	26.3	.2	7.5	11.55	1.45	19.98	15.19	87.59
.59	102.6	18.1	7.5	25.6									
					17.6	25.0	.2	7.6	11.70	1.75	20.08	16.06	130.7
.57	95.0	17.0	7.3	24.3									
					16.5	23.8	.1	8.1	12.47	1.04	23.78	19.98	123.6
.55	86.9	16.0	7.2	23.2									
					15.6	22.7	.2	8.1	12.47	2.47	21.17	18.66	142.3
.53	78.8	15.1	7.0	22.1									
					14.7	21.6	.1	7.3	11.24	1.42	21.20	19.63	161.9
.51	71.5	14.2	6.9	21.1									
					13.8	20.7	.1	6.1	9.39	1.62	17.48	16.89	178.8
.49	65.4	13.4	6.8	20.2									
					12.7	19.4	.3	8.1	12.47	5.36	16.59	17.10	195.9
.45	57.3	12.0	6.5	18.5									
					11.2	17.6	.3	5.6	8.62	5.88	6.77	7.69	203.6
.40	51.7	10.4	6.2	16.6									
					9.8	15.9	.2	4.3	6.62	4.07	6.71	8.44	212.0
.35	47.4	9.1	6.0	15.1									
					8.5	14.4	.3	4.4	3.70	2.53	3.36	4.66	220.5
.30	45.0	7.9	5.7	13.6									

[a] $\Sigma V_p = 0.29\ cm^3\ g^{-1}$.
[b] $\Sigma S = 220.5\ m^2\ g^{-1}$.

multilayer of adsorbed film of depth t exists on the pore wall when evaporation or condensation occurs which is of the same depth as the adsorbed film on a nonporous surface.

The method shown in Table 9-1 uses data from either the adsorption or desorption isotherm. However, as stated previously, the desorption curve is usually employed except in those cases where the adsorption curve corresponds to the thermodynamically more stable condition such as occurs in the case of "bottleneck" pores. In either case, for case of presentation, the data is evaluated downward from high to low relative pressures.

Columns 1 and 2 of Table 9-1 contain data obtained directly from the isotherm. The adsorbed volumes are normalized for one gram of adsorbent. Relative pressures are chosen using small decrements at high values, where r_k is very sensitive to small changes in relative pressure and where the slope of the isotherm is large, such that small changes in relative pressure produce a large change in volume.

Column 3, the Kelvin radius, is calculated from the Kelvin equation assuming a zero wetting angle. If nitrogen is the adsorbate, equation [9.18] can be used.

Column 4, the film depth t, is calculated using equation [9.25], the Halsey equation.

Column 5 is the pore radius obtained from equation [9.19].

Columns 6 and 7, $\bar{r}_k$ and $\bar{r}_p$, are prepared by calculating the mean value in each decrement from successive entries.

Column 8, the change in film depth, is calculated by taking the difference between successive values of t.

Column 9, ΔV_{gas}, is the change in adsorbed volume between successive P/P_0 values and is determined by subtracting successive values from column 2.

Column 10, ΔV_{liq}, is the volume of liquid corresponding to ΔV_{gas}. The most direct way to convert ΔV_{gas} to ΔV_{liq} is to calculate the moles of gas, and multiply by the liquid molar volume. For nitrogen at standard temperature and pressure this is given by

$$\Delta V_{liq} = \frac{\Delta V_{gas}}{22.4 \times 10^3} \times 34.6 = \Delta V_{gas}(1.54 \times 10^{-3}) \text{ cm}^3 \quad [9.26]$$

Column 11 represents the volume change of the adsorbed film remaining on the walls of the pores from which the center core has previously evaporated. This volume is the product of the film area ΣS and the decrease in the film depth Δt. By assuming no pores are present larger than 950 Å ($P/P_0=0.99$) the first entry in column 11 is zero since there exists no film area from previously emptied pores. The error this assumption introduces is negligible because the area produced by pores larger than 950 Å will be small compared to their volume.

Subsequent entries in column 11 are calculated as the product of Δt for a decrement and ΣS from the row above corresponding to the adsorbed film area exposed by evaporation of the center cores during all the previous decrements.

Column 12, the actual pore volume, is evaluated by recalling that the volume of liquid, column 10, is composed of the volume evaporated out of the center cores plus the volume desorbed from the film left on the pore walls. Then,

$$\Delta V_{\text{liq}}=\pi \bar{r}_k^2 L+\Delta t \Sigma S \qquad [9.27]$$

and since,

$$V_p=\pi \bar{r}_p^2 L \qquad [9.28]$$

then, by combining the above preceding equations,

$$V_p=\left(\frac{\bar{r}_p}{\bar{r}_k}\right)^2 [\Delta V_{\text{liq}}-(\Delta t \Sigma S)(10^{-4})]\ \text{cm}^3 \qquad [9.29]$$

Column 13 is the surface area of the pore walls calculated from the pore volume by

$$S=\frac{2V_p}{\bar{r}_p}\times 10^4\ (\text{m}^2) \qquad [9.30]$$

with V_p in cubic centimeters and $\bar{r}_p$ in Angstroms. It is this value of S which is summed in column 14. The summation is multiplied by Δt from the following decrement to calculate the film volume decrease in column 11.

If the utmost rigor were used, the summation of areas from the center core should be employed to calculate the volume of desorbed film since it is this area times the change in film depth which determines the volume change of the film during each decrement. However, the surface area of the adsorbed film changes with each decrement and it would be cumbersome to account for the change in core area of previously emptied pores with each successive decrement. Furthermore, any attempt to improve the accuracy in this manner is questionable in view of the many other assumptions which have been made.

Table 9-1 discloses that the volume of all pores greater than 13.6 Å is 0.29 $cm^3\ g^{-1}$. This does not mean that micropores with smaller radii are absent but rather, as stated earlier, that the validity of the Kelvin equation becomes questionable because of the uncertainty regarding molar volumes and surface tension when only one or two molecular diameters are involved.[65-66] Termination of the analysis is necessary, therefore, when the relative pressure approaches 0.3. In some analyses it is found that the volume desorbed from the film ($\Delta t \Sigma S$) becomes equal to the total amount desorbed, indicating that only desorption is occurring in the narrower pores and the analysis must be ended because in the absence of evaporation from center cores the Kelvin equation is not applicable.

The volume of micropores, if present, can be evaluated by the difference between the total pore volume (see equation [9.15]) and the sum of column 12, ΣV_p.

The total area of all pores to 13.6 Å radius (column 14) is 220.5 $m^2\ g^{-1}$. This area is usually less than the BET area since it does not include the surface contributed by micropores.

An implicit assumption, hidden in the method, is that no surface exists other than within pores because the volume desorbed in any decrement is assumed to originate from either the center core or the adsorbed film on the pore wall.

Figure 9-4 is the pore size distribution curve prepared from the data contained in Table 9-1. The terms ΔV_p and Δr_p are taken from columns 12 and 5, respectively. Values of r_p are from column 7. Figure 9-5 is the same data plotted as the cumulative pore volume.

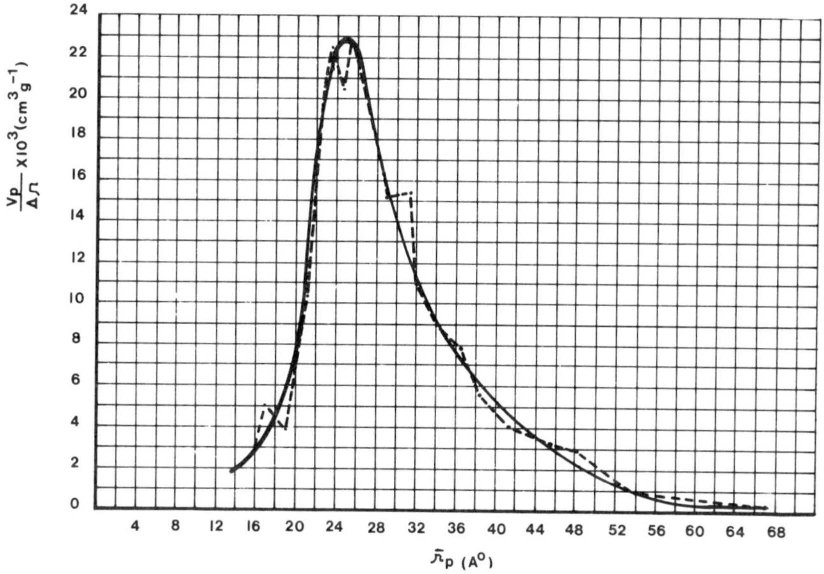

Figure 9-4 Pore size distribution curve from Table 9-1.

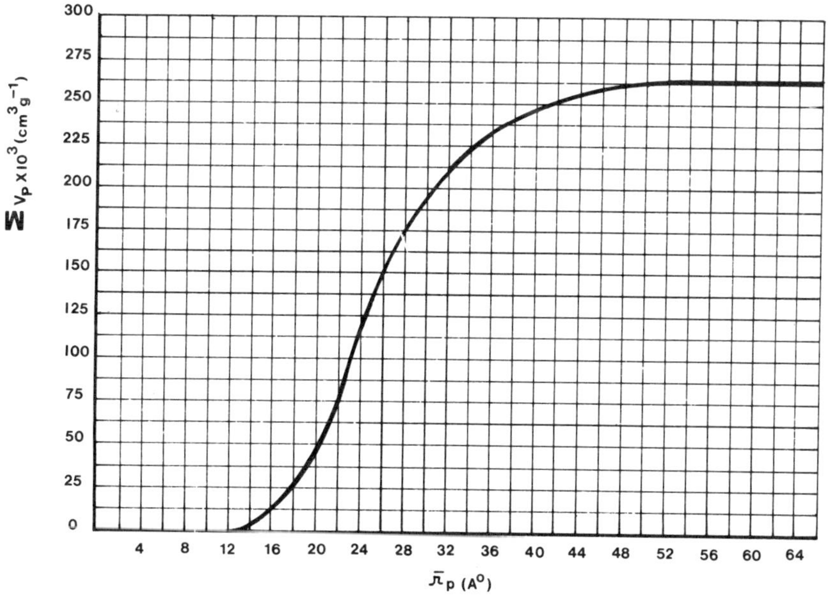

Figure 9-5 Cumulative pore volume from Table 9-1.

6. Modelless Pore Size Analysis

The ideas of Wheeler[64] that condensation and evaporation occur within a center core during adsorption and desorption and that an adsorbed film is present on the pore wall has led to the proposal of various methods for pore size analysis. In addition to the methods of Pierce[58] and the BJH[63] technique, other schemes have been proposed, including those by Shull,[56] Oulton,[67] Roberts,[68] Innes,[69] and Cranston and Inkley.[60] These ideas are all based upon some assumption regarding the pore shape.

Brunauer[70] et al. have developed a means of determining the pore volume distribution wherein the pore shape has a negligible influence.

Equation [9.4] establishes the relationship between the moles condensed into pores and the corresponding decrease in the pore area. Rewriting equation [9.4] and eliminating the negative sign, since during condensation pore surface is disappearing, one obtains

$$S=\frac{1}{\gamma}\int\left(RT\ln\frac{P}{P_0}\right)dn \qquad [9.31]$$

Kiselev,[71] using the preceding equation by graphical integration of the isotherm between the limits of saturation and hysteresis loop closure, was able to calculate surface areas for wide pore samples in good agreement with BET measured areas. For micropores, the absence of hysteresis at the low-pressure end of the isotherm indicates that only adsorption and not condensation occurs, thereby rendering Kiselev's method inapplicable.

Brunauer's modelless method uses pore volume and pore area not as functions of the Kelvin radius but rather as functions of hydraulic radii that he defines as

$$r_h=\frac{V}{S} \qquad [9.32]$$

where V and S are the pore volume and surface, respectively, regardless of their shape. For cylinders and parallel plates r_h is one-half the radius or distance between plates.

As in the method of Pierce,[58] Brunauer assumes that in the first decrement, say between $P/P_0=1$ and 0.95, desorption occurs only from the center core of the largest pores regardless of their shape. The hydraulic radius of the core is calculated by dividing the liquid volume which has evaporated out of the pore by the core area as determined by graphically integrating equation [9.31] using $P/P_0=1$ and 0.95 as the limits of integration.

In the second decrement the liquid volume desorbed $(V_{\text{liq}})_2$ must be corrected for the decrease in the adsorbed film depth remaining on the walls of previously emptied pores. By assuming the pores are cylindrical (an unnecessary assumption), the core volume $(V_k)_2$ can be calculated as

$$(V_k)_2=(V_{\text{liq}})_2-S_1(t_1-t_2)-\frac{S_1}{4(r_h)_1}(t_1-t_2)^2 \qquad [9.33]$$

where S_1 and $(r_h)_1$ are the core surface and hydraulic radius calculated from the first decrement. The terms t_1 and t_2 refer to the adsorbed film depth at the beginning and end of the second decrements. These values can be obtained from equation [9.25], the Halsey equation. The volume $(V_k)_2$ is divided by S_2 calculated from equation [9.31] using as limits for the graphical integration the relative pressures at the beginning and end of the second decrement. This ratio gives the second hydraulic radius $(r_h)_2$.

To calculate the hydraulic radius in the third decrement, the desorbed volume must be corrected for the contribution made by the film on the walls of pores emptied in the first decrement and decreased in depth during the second accrement plus the contribution made by the change in film depth within pores emptied in the second decrement. The total correction factor is

$$(V_k)_3=(V_{\text{liq}})_3-S_1(t_1-t_3)-\frac{S_1}{4(r_h)_1}(t_1-t_3)^2-S_1(t_1-t_2)$$
$$-\frac{S_1}{4(r_h)_1}(t_1-t_2)^2-S_2(t_2-t_3)-\frac{S_2}{4(r_h)_2}(t_2-t_3)^2 \qquad [9.34]$$

It becomes painfully and quickly obvious that for an analysis requiring many data points, the number of correction terms becomes cumbersome. However, Brunauer was able to show that the squared terms do not contribute significantly and the preceding equation can be rewritten as

$$(V_k)_3 = (V_{liq})_3 - S_1(t_1 - t_3) - S_1(t_1 - t_2) - S_2(t_2 - t_3) \quad [9.35]$$

Equation [9.35] is identical to the correction factor for parallel plate pores.

Using the preceding corrections, the Brunauer method is not modelless. However, it does offer a means of employing the same correction factor for pores as diverse as parallel plates and cylinders. In those instances where the pore geometry does differ considerably from cylinders or parallel plates the error introduced by assuming either of these shapes is small. This was confirmed by Brunauer by plotting pore distributions (V_k/r_h) versus r_h using corrected and uncorrected core volumes. These plots differed only slightly, and the hydraulic radius remained essentially unchanged. Accordingly, if uncorrected hydraulic core volumes are used the method is entirely modelless and little accuracy is sacrificed.

7. *V–t* Curves

In Chapter 5, it was emphasized that the surface of an adsorbent is never covered with an adsorbed film of uniform depth but rather with stacks of adsorbed molecules of various depths. Discussions in this chapter regarding film depth t on the pore walls implies that the film is uniform. To be consistent with the BET theory, the adsorbed film on the wall of a pore is structured as on a nonporous surface and the film depth t is in reality a "statistical depth" that is proportional to the number of monolayers present in the film regardless of how the molecules may be stacked.

Shull[56] showed that on a number of nonporous solids, the ratio of the weight adsorbed W_a and the weight corresponding to formation of a monolayer W_m when plotted versus the relative pressure could be closely represented by a single curve regardless of the solid. The curve produced by Shull is a typical type II isotherm. Similar plots were made by others[55–60] on a variety of nonporous materials and in each case the points reasonably well fit the same common curve, particularly at relative pressures above 0.3. The common curve is also fit by the Halsey[61] equation above 0.3 relative pressure. If the monolayer is envisioned as being uniformly one molecule in depth, then a plot of W_a/W_m versus P/P_0 discloses the relative pressures corresponding to surface coverage by any number of monolayers. Therefore, if the adsorbate diameter is known, the statistical depth t can be calculated by multiplying the number of monolayers by the adsorbate diameter. Figure 9-6 is a t plot prepared from the Halsey equation. Experimental t plots lie slightly above and below the indicated curve.

Shull[56] assumed that the adsorbate molecules were packed one on top of the other in the film and deduced the monolayer depth to be 4.3 Å for nitrogen. A more realistic assumption is that the film structure is close packed hexagonal, leading to a monolayer depth of 3.54 Å as shown by equation [9.23]. Therefore, as stated previously, the statistical depth of the adsorbed film is

$$t = \left(\frac{W_a}{W_m}\right) 3.54\ \text{Å} \qquad \text{[cf. 9.24]}$$

If the volume adsorbed, on a surface of area S, is expressed as the corresponding liquid volume, then

$$t = \frac{V_{\text{liq}}}{S} \times 10^4\ (\text{Å}) \qquad [9.36]$$

with V_{liq} in cubic centimeters and S in square meters.

Lippens and deBoer[57] have shown that a plot of the volume adsorbed V_{liq} versus t calculated from equation [9.24] will yield a straight line of slope $S \times 10^{-4}$ from which the surface area can be calculated. Plots of this nature are termed *V–t* curves and generally give surface areas comparable to BET values provided that multilayer adsorption and not condensation into the pores takes place.

Figure 9-7 illustrates three types of *V–t* curves. Curve X results from multilayer adsorption in the absence of any condensation into pores. Curve Y shows that condensation into pores commences where the slope starts to increase. The upward deviation from a straight line reflects the increased uptake due to condensation in larger pores. Curve Z results when narrow pores are filled at low

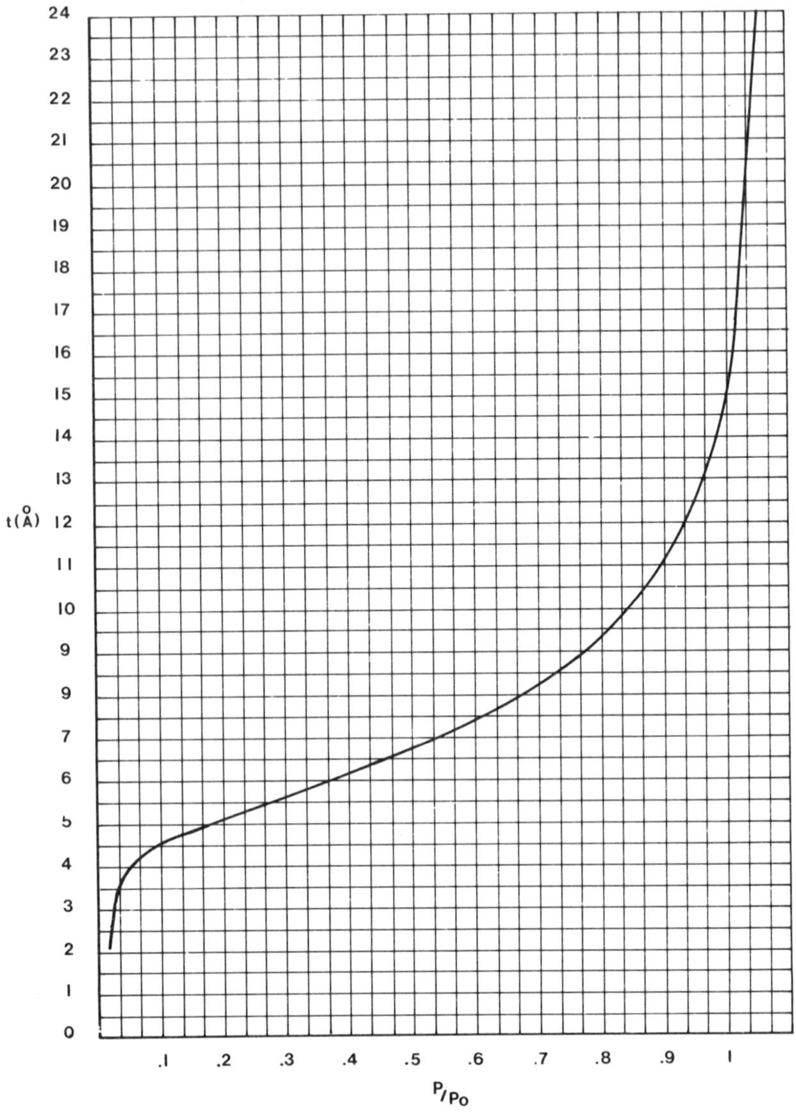

Figure 9-6 *t*-Plot derived from the Halsey equation.

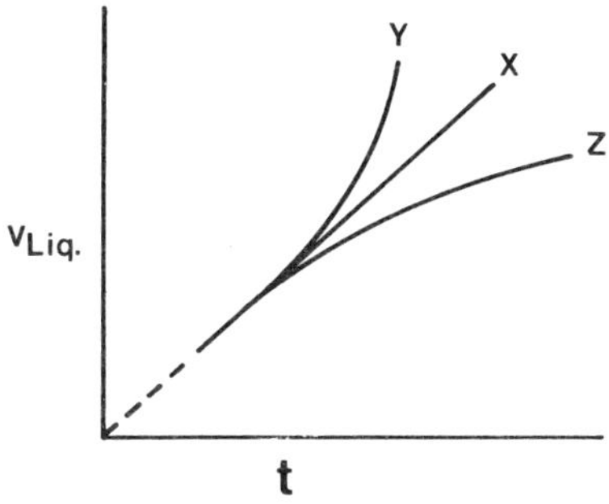

Figure 9-7 Basic shapes, *V–t* curves.

relative pressures by multilayer adsorption thereby reducing the available surface for continued adsorption.

CHAPTER TEN

Microporosity

1. Introduction

It is convenient to categorize pores into several size ranges. The largest pore radii amenable to analysis by the Kelvin equation are about 1000 Å, which corresponds to relative pressures near 0.99. Pores with radii greater than this are labeled "macropores." Dubinin[72] calls pores in the Kelvin range, from about 15 to 1000 Å, "transitional" and pores with radii less than about 15 Å, "micropores." The failure of *V–t* plots to pass through the origin has led to the postulation of "submicropores"[73] with radii less than about 7 Å.

Microporous materials exhibit type I isotherms which are described by equation [5.12], the Langmuir equation. The Langmuir model was developed on the assumption that adsorption was limited to at most one monolayer. However, any factor that can limit the quantity adsorbed to a few monolayers will also produce a type I isotherm. In the case of very narrow pores, the close proximity of the walls will prevent multilayer adsorption and limit the amount adsorbed.

In view of the intense potential fields in very narrow pores it would be difficult to determine if the mechanism by which the pores fill is that of adsorption or condensation. However, equation [5.39], a form of the BET equation that treats restricted adsorption, allows for just this kind of limiting adsorption.

2. Langmuir Plots for Microporous Surface Area

A convenient form of the Langmuir equation is

$$\frac{P}{W} = \frac{1}{KW_m} + \frac{P}{W_m} \qquad \text{[cf. 5.12]}$$

where P is the adsorbate equilibrium pressure, and W and W_m are the weight adsorbed and monolayer weights, respectively. The term K is a constant given in Chapter 5, Part 1.

For type I isotherms, a plot of P/W versus P should give a straight line with $1/W_m$ as the slope. The sample surface area S_t is calculated from equation [5.13]:

$$S_t = \frac{W_m \overline{N}}{\overline{M}} \mathscr{A} \qquad \text{[cf. 5.13]}$$

where, as before, $\mathscr{A}$ is the cross-sectional adsorbate area, $\overline{M}$ is the adsorbate molecular weight, and $\overline{N}$ is Avogadro's number.

The fact that a Langmuir plot gives a straight line is not at all indicative of its success. Without an understanding of the processes occurring within the micropores in terms of adsorption or condensation, the Langmuir equation may be a correct mathematical description of the isotherm but the interpretation of the meaning of the monolayer weight W_m remains unresolved.

3. Extensions of Polanyi's Theory for Micropore Volume and Area

Polanyi's[74] potential theory of adsorption views the area immediately above an adsorbent's surface as containing equipotential lines which follow the contour of the surface potential. When a molecule is adsorbed it is considered trapped between the surface and the limiting equipotential plane at which the "adsorption potential" has fallen to zero. Figure 10-1 illustrates the equipotential planes discussed previously. In the diagram, Y represents a pore and X depicts some surface impurity.

According to the potential theory the volume $\tilde{V}$, defined by the

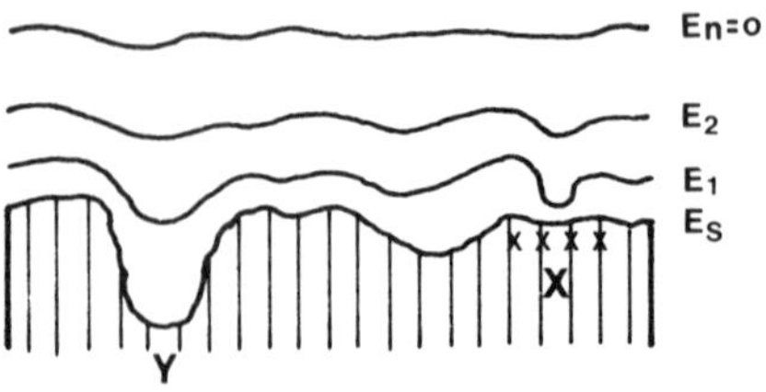

Figure 10-1 Polanyi's potential planes.

adsorbent's surface and the equipotential plane E_n, can contain adsorbate in three different conditions depending upon the temperature. Above the critical temperature the adsorbate cannot be liquified, and the gas in the adsorption volume $\tilde{V}$ simply becomes more dense near the surface. At temperatures near but less than the critical temperature, the adsorbate is viewed as a liquid near the surface and a vapor of decreasing density away from the surface. Substantially below the critical temperature ($T \leqslant 0.8 T_c$), the adsorption volume is considered to contain only liquid. Under the latter conditions one can write

$$\tilde{V} = \frac{W}{\rho} \qquad [10.1]$$

where W and ρ are the adsorbate weight and density, respectively.

The potential theory asserts that the adsorption potential, when the adsorbate is in the liquid state, is given by

$$E = RT \ln \frac{P_0}{P} \qquad [10.2]$$

According to the preceding equation, E is the isothermal work required to compress the vapor from its equilibrium pressure P to the saturated pressure P_0 of the liquid in the adsorption volume.

With equations [10.1] and [10.2] both $\tilde{V}$ and E can be calculated from an experimental isotherm. Therefore,

$$\tilde{V} = F(E) \qquad [10.3]$$

Plots of $\tilde{V}$ versus E take the form shown in Figure (10-2) and are called characteristic curves.

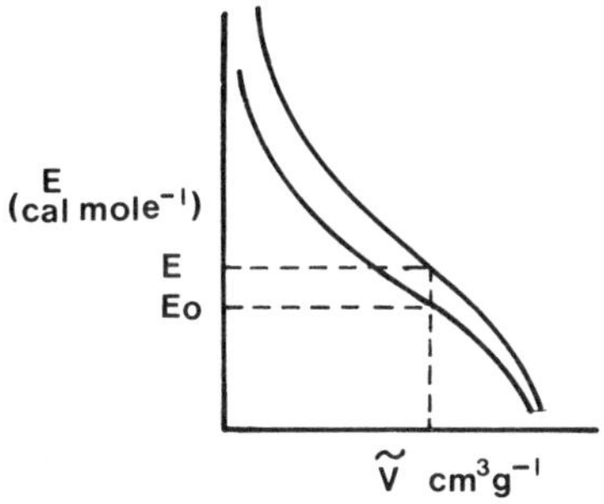

Figure 10-2 Characteristic curves for two vapors.

If two adsorbates fill the same adsorption volume as shown by the vertical dotted line in Figure 10-2, their adsorption potentials E and E_0 will differ only because of differences in their molecular properties. Consequently, the ratio of adsorption potentials is assumed by Dubinin[75] to be constant and he calls E/E_0 the "affinity coefficient" which, for an adsorbate pair, is a measure of their relative affinities for a surface or their adsorbability. Using the adsorption for one vapor as a reference value, say E_0, the ratio of potentials can be written as

$$\frac{E}{E_0}=\beta \qquad [10.4]$$

Substitution into equation [10.3] then gives, for the reference vapor,

$$\tilde{V}=F\left(\frac{E}{\beta}\right) \qquad [10.5]$$

Using benzene as the reference or standard vapor for which β is taken as unity, Dubinin and Timofeev[76] were able to calculate values of β for other adsorbents.

The similarity in the shapes of the characteristic curves, shown in Figure 10-2, and the positive side of a Gaussian curve, led Dubinin and Radushkevich[77] to postulate that the fraction of the adsorption volume $\tilde{V}$ occupied by liquid adsorbate at various values of adsorption potentials E can be expressed as a Gaussian function. Thus,

$$\tilde{V} = \tilde{V}_0 e^{-KE_0^2} \quad [10.6]$$

Substituting the value for E_0 from equation [10.4] gives

$$\tilde{V} = \tilde{V}_0 e^{-K(E/\beta)^2} \quad [10.7]$$

where K is a constant determined by the shape of the pore size distribution curve and $\tilde{V}$ is the total adsorption volume or the total microporous volume.

Equation [10.7] is applicable to micropores rather than larger pores because the overlapping potential from the walls of pores only slightly larger than an adsorbate molecule will considerably enhance the adsorption potential.

Substituting for E in equation [10.7] using equation [10.2] yields

$$\tilde{V} = \tilde{V} \exp - \left[\frac{K}{\beta^2}\left(RT \ln \frac{P_0}{P}\right)^2\right] \quad [10.8]$$

which can be rewritten as

$$\log W = \log(\tilde{V}_0 \rho) - 2.303\, K \left(\frac{RT}{\beta}\right)^2 \left[\log\left(\frac{P_0}{P}\right)\right]^2 \quad [10.9]$$

where W and ρ are the weight adsorbed and the liquid adsorbate density.

Simplification of equation [10.9] gives

$$\log W = \log(V_0 \rho) - k\left[\log\left(\frac{P_0}{P}\right)\right]^2 \quad [10.10]$$

where

$$k = 2.303\, K \left(\frac{RT}{\beta}\right)^2 \quad [10.11]$$

A plot of log W versus log $[(P_0/P)]^2$ should give a straight line with an intercept of log $(\tilde{V}_0\rho)$ from which $\tilde{V}_0$, the micropore volume, can be calculated. Nikolayev and Dubinin[78] found linear plots using relative pressures ranging from 10^{-5} to 10^{-1} on a variety of microporous samples.

Kaganer[79] modified Dubinin's method in order to calculate the surface area within micropores. He assumed that the adsorption potential of the sites is distributed according to a Gaussian function such that

$$\theta = e^{-\bar{k}E^2} \tag{10.12}$$

where θ is the fraction of the surface covered by adsorbate. Combining equations [10.12] and [10.2] in logarithmic form yields

$$\log \theta = -2.303\bar{k}(RT)^2 \left(\log \frac{P_0}{P}\right)^2 \tag{10.13}$$

The fraction of surface θ, covered by adsorbate, for less than monolayer coverage, is the adsorbed weight W relative to the monolayer weight W_m or

$$\theta = \frac{W}{W_m} \tag{10.14}$$

Equation [10.13] then becomes

$$\log W = \log W_m - \bar{K}\left(\log \frac{P_0}{P}\right)^2 \tag{10.15}$$

where

$$\bar{K} = 2.303\bar{k}(RT)^2 \tag{10.16}$$

Equation [10.15] is similar to equation [10.10], Dubinin's equation. A plot of log W versus $(\log P_0/P)^2$ will yield a straight line with an intercept of log W_m from which the surface area can be calculated by equation [5.13]. The linear range of these plots are usually at very low relative pressures, less than 10^{-2}. Kaganer showed excellent agreement between surface areas measured using equation [10.15] and the BET method on a variety of adsorbate–adsorbent systems.

4. The MP Method

Mekhail, Brunauer, and Bodor[80] have devised a method for the analysis of micropores which offers several advantages. These include the ability to obtain the micropore volume, surface, and their distributions from one experimental isotherm. Data for the MP (micropore analysis) method need not be measured at the very low pressures needed for the Dubinin and Kaganer theories. The method assumes no site energy or adsorption volume distribution functions as required by the Kaganer and Dubinin theories. Also, the MP method is applicable to adsorbents containing macro-, transitional (mesopores), and micropores with self-termination when the micropore analysis is completed.

When measuring mesopores, the statistical thickness t is used as a correction to allow for desorption from the adsorbed film. However, the value of t is much more critical when measuring micropores, because in the MP method, t is the actual measure of the pore size.

The usual method of measuring the statistical thickness is to divide the liquid volume of nitrogen absorbed, by the BET surface area

$$t = \frac{V_{\mathrm{liq}}}{S_{\mathrm{BET}}} \times 10^4 \ (\text{Å}) \qquad [\text{cf. } 9.36]$$

The assumption usually made is that the ratio $V_{\mathrm{liq}}/S_{\mathrm{BET}}$ has the same value at a given relative pressure independent of the solid. A plot therefore of t versus P/P_0 should give the same curve for any nonporous solid, (see Figure 9-6). In fact, plots of the number of adsorbed layers versus P/P_0 show some discrepancies which for the analysis of large pores is not significant. Therefore, the Halsey equation can be used for the statistical thickness in that application. However, for micropore analysis, a statistical thickness must be taken from a t versus P/P_0 curve that has approximately the same BET C value as the test sample. The unavailability of t versus P/P_0 plots on numerous surfaces with various C values would make the MP method of passing interest were it not for the fact that t can be calculated from equation [9.36]. This implies that surface area can be accurately measured on microporous samples. Brunauer[81] points out that in most instances the BET equation does correctly measure the micropore surface area.

As stated earlier the C value in micropores will be large due to the overlapping wall potentials. Under these circumstances, the surface will be covered well over 90% by stacks of adsorbate not in excess of two molecules in depth as shown by equation [5.45] and Table 5-1. Therefore, the close proximity of the walls offers no special condition that is not already allowed for by the BET theory.

To illustrate the MP method, consider the isotherm shown in Figure 10-3. The volumes of adsorbed gas are converted to liquid volumes from which t is calculated using equation [9.36]. The V_t plot shown in Figure 10-4 is then constructed using relative pressure intervals of 0.05 starting at 0.05.

In the example shown the t values were not calculated from equation [9.36] but were taken from a t versus P/P_0 plot prepared by deBoer[82] on a sample with a similar BET C value. By choosing t values from a material with a similar but not identical C value the surface area nevertheless agreed to within 1.4% of the BET measured area.

According to Figure 9-7, it is evident that the shape of Figure 10-4 is to be associated with microporosity. The slope of the linear

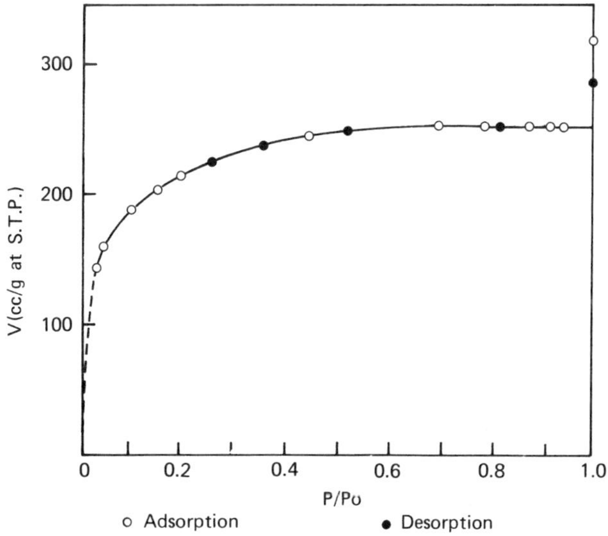

Figure 10-3 Isotherm of N_2 on silica gel, Davidson 03 at 77.3°K (from reference 80).

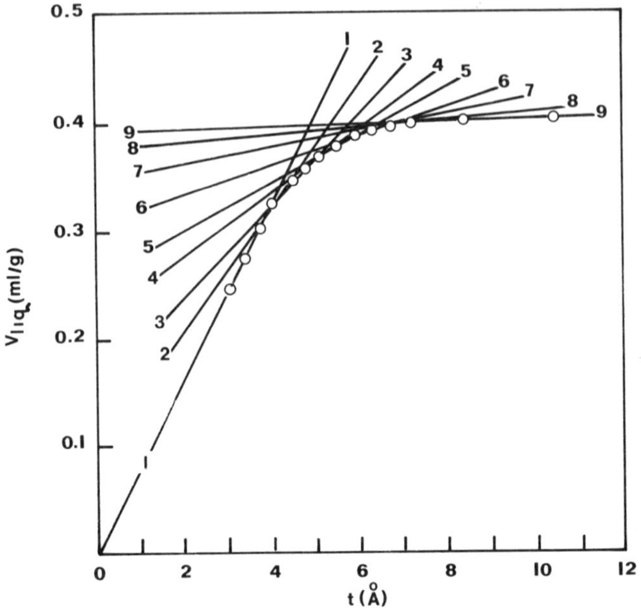

Figure 10-4 *V–t* curve from Figure 10-3 (from reference 80).

portion of the curve, the origin through the first four points, is 0.0792 which gives a micropore surface of 792 m^2 when calculated from equation [9.36]. Straight line 2, drawn tangent to the curve between $t=4$ and 4.5 Å, exhibits a slope of 0.0520. The area of all the pores remaining unfilled by adsorbate is 520 m^2 and the surface area of pores in the range of thickness from 4 to 4.5 Å is $792-520=272$ m^2. The third line gives a slope of 0.0360 between t values of 4.5 and 5 Å and the area of pores in this range is $520-360=160$ m^2. The calculation is continued in this manner until there is no further decrease in the slope of the *V–t* plot which means that all the micropores are filled.

The calculation of pore volume is carried out in an equally straightforward manner. For example, the volume of pores is given by

$$V=10^{-4}(S_1-S_2)\frac{(t_1+t_2)}{2}\ \text{cm}^3\ \text{g}^{-1} \qquad [10.17]$$

Thus, for the first group of pores, the volume is

$$V = 10^{-4}(792 - 520)(4.0 + 4.5) = 0.1156\ \text{cm}^3\ \text{g}^{-1} \quad [10.18]$$

Table 10-1 contains the micropore data for the isotherm shown in Figure 10-3.

Table 10-1.

Pore group	S_{i+1} (m^2/g)	$S_i - S_{i+1}$ (m^2/g)	r_h (Å)	V_i (cm^3/g)
1	520	272	4.25	0.1156
2	360	160	4.75	0.0760
3	280	80	5.25	0.0420
4	200	80	5.75	0.0460
5	140	60	6.25	0.0375
6	80	60	6.75	0.0405
7	20	60	7.25	0.0435
8	10	10	7.75	0.0077
		$\Sigma S_i = 782$		

BET area $= 793\ m^2\ g^{-1}$
V–t area $782\ m^2\ g^{-1}$
Total pore volume $= .4034$
MP pore volume $= .4088$

The authors attributed the difference between the V–t and BET areas to surfaces that did not lie within pores
x

The exact pore shape is usually unknown and cylindrical pores are generally assumed. Mikhail, Brunauer, and Bodor[80] show in their paper that equation [10.17] is equally valid for parallel plate or cylindrical pores and that the hydraulic radius r_h, in Table 10-1, is the same as the separation between plates or the cylindrical radius.

CHAPTER ELEVEN

Mercury Porosimetry

1. Calculation of Pore Radii

The surface tension γ of a liquid is defined as the work required to produce one square centimeter of surface. If a narrow capillary is immersed in a liquid, as shown in Figure 11-1, the liquid within the capillary will rise above, or fall below, the external liquid level depending upon the liquid contact or wetting angle θ with the solid surface.

Because of mercury's high surface tension it tends not to wet* most solid surfaces and must be forced to enter a pore. When forced under pressure into a pore of radius r and length l an amount of work W is required which is proportional to the increased surface exposed by the mercury at the pore wall. Therefore, assuming cylindrical pores

$$W = 2\pi r l \gamma \qquad [11.1]$$

Since mercury exhibits a wetting angle greater than 90° and less than 180° on all surfaces with which it does not amalgamate, the

*Complete wetting occurs at a contact angle of zero degrees and total nonwetting at 180°. Wetting and nonwetting are loose terms usually associated with contact angles less than and greater than 90°, respectively.

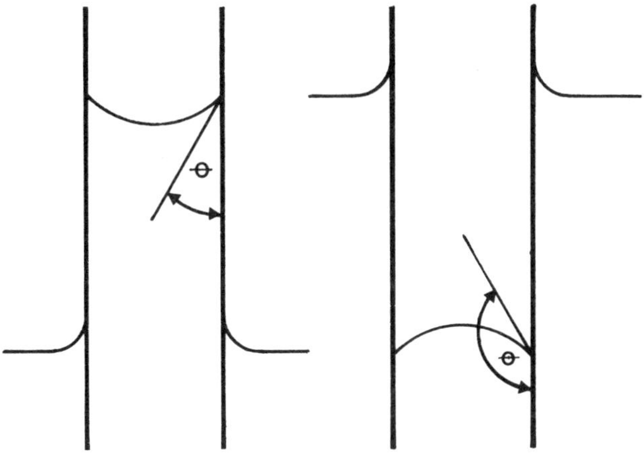

Figure 11-1 Left: Capillary rise; liquid wets the solid; $\theta < 90^\circ$. Right: Capillary depression; liquid does not wet solid; $\theta < 90^\circ$.

work required is reduced by cos θ and equation [11.1] becomes

$$W_1 = 2\pi r l \gamma \cos\theta \qquad [11.2]$$

When a volume of mercury ΔV is forced into a pore under external pressure P an amount of work W_2, given as follows, is performed.

$$W_2 = P\Delta V = P\pi r^2 l \qquad [11.3]$$

At equilibrium, equations [11.2] and [11.3] are combined to give

$$Pr = 2\gamma \cos\theta \qquad [11.4]$$

Because the product Pr is constant, assuming constancy of γ and θ, equation [11.4] dictates that as the pressure increases, mercury will intrude into progressively narrower pores. Equation [11.4] was developed by Washburn.[83]

The volume forced into the pores is usually monitored in a penetrometer, which is the calibrated stem of a glass cell containing the sample and filled with mercury. As intrusion occurs the level in the capillary stem decreases. Ritter and Drake[84] used a resistance wire coaxial in the capillary stem to monitor the volume changes. They also developed one of the earliest high-pressure porosimeters and measured the contact angle between mercury and a variety of materials, which they found to be between 135 and 142° with an average value of 140°. Using this value for θ and 484 erg cm^{-2}, at room temperature, for γ, converts Equation [11.4] into the following equations using pressure in pounds per square inch absolute (psia) or kilograms per square centimeter.

$$r = \frac{106.7}{P}; \ P \text{ in psia, } r \text{ in micrometers} \qquad [11.5]$$

and

$$r = \frac{1530}{P}; \ P \text{ in kg cm}^{-2}, r \text{ in micrometers} \qquad [11.6]$$

2. Surface Area From Intrusion Curves

Figure 11-2 is a cumulative pore volume curve which shows the summation of volume intruded into the pores and interparticle voids versus the applied pressure. The surface area of all pores and voids filled up to pressure P is calculated using equations [11.2] and [11.3] as follows:

$$2\pi r l \gamma |\cos\theta| = P\Delta V \qquad [11.7]$$

Assuming cylindrical pores open at each end and recognizing that $2\pi rl$ is the pore area S, gives

$$S\gamma|\cos\theta| = P\Delta V \qquad [11.8]$$

Then, pores in the radius range dr which take up volume dV will possess area dS as given by

$$dS = \frac{P\,dV}{\gamma|\cos\theta|} \qquad [11.9]$$

The area of pores filled up to radius r is given by integration of the preceding equation which is, assuming constancy of surface tension and wetting angle, given as

$$S = \frac{1}{\gamma|\cos\theta|}\int_0^V P\,dV \qquad [11.10]$$

For various pressure units with V in cubic centimeters, $\theta=140^\circ$, and $\gamma=484$ erg cm^{-2}, equation [11.10] becomes

$$S=0.0186\int_0^v PdV\ (\mathrm{m}^2);\ P=\mathrm{psia} \qquad [11.11]$$

$$S=2.645\times10^{-5}\int_0^v PdV\ (\mathrm{m}^2);\ P=\mathrm{kg\ m}^{-2} \qquad [11.12]$$

By graphical integration, using the cumulative curve of Figure 11-2 the surface area of all pores filled by mercury up to 60,000 psia (0.00179 μm) is calculated as follows:

A. Area above the total curve to the maximum intruded volume measured by planimetry $=55.2$ cm^2

[11.13]

B. $$\text{Normalization factor}=\frac{(10^4\ \mathrm{psia})(10^{-1}\ \mathrm{cm}^3\ \mathrm{g}^{-1})}{4\ \mathrm{cm}^2\ \text{of graph paper}}$$

$$=250\ \frac{\mathrm{psia\ cm}^3\ \mathrm{g}^{-1}}{\mathrm{cm}^2} \qquad [11.14]$$

C. $$\int_0^v PdV=(55.2\ \mathrm{cm}^2)\frac{(250\ \mathrm{psia\ cm}^3\ \mathrm{g}^{-1})}{\mathrm{cm}^2}$$

$$=13{,}800\ \mathrm{psia\ cm}^3\ \mathrm{g}^{-1} \qquad [11.15]$$

D. $$S=0.0186\int_0^v PdV=(0.0186)(46{,}850)=256.7\ \mathrm{m}^2\ \mathrm{g}^{-1}$$

[11.16]

The surface area of pores in a radius interval can be obtained by the preceding method. For example, the area W over the curve between 5000 psia ($r=0.0215$ μm) and 60,000 psia ($r=0.00179$ μm) is 50.1 cm^2; then

$$S=0.0186\times50.1\times250=232.9\ \mathrm{m}^2 \qquad [11.17]$$

Therefore, the surface area of pores in the range 0.0215–0.00179 μm is 232.9 m^2. The area of the large pores and interparticle voids is given by the difference between the total area and the area above or 23.8 m^2.

The high initial slope of the cumulative curve (Figure 11-1) is attributed to intrusion into the large pores and interstices between particles. The upward inflection near 5000 psia indicates that pores of about 0.02 μm are commencing to fill and that intrusion occurs into progressively smaller pores as the pressure is raised. Above 30,000 psia no appreciable additional intrusion takes place indicating that little pore volume exists in pores narrower than 0.0035 μm.

3. Pore Size Distribution

Allow dV to be the volume element of all pores with radii between r and $r+dr$. Then

$$dV = D_v(r)dr \qquad [11.18]$$

where $D_v(r)$ is the volume pore size distribution function defined as the volume per unit interval of radius.

Differentiation of equation [11.4], assuming constancy of γ and θ, yields

$$Pdr + rdP = 0 \qquad [11.19]$$

Combining equations [11.18] and [11.19] gives

$$-dV = D_v(r)\frac{r}{P}\,dP \qquad [11.20]$$

Then

$$D_v(r) = \frac{-P}{r}\left(\frac{dV}{dP}\right) \qquad [11.21]$$

Since the change in volume is measured as a decreasing volume the negative sign can be eliminated to give

$$D_v(r) = \frac{P}{r}\left(\frac{dV}{dP}\right) \qquad [11.22]$$

Equation [11.21] represents a convenient means of reducing the cumulative curve to the distribution curve which gives the pore volume per unit radius interval.

Table 11-1. Data and Calculations for the Cumulative (Fig. 11-2) and Distribution (Fig. 11-3) Curves

P (psia)	ΔP (psia)	$r \times 10^{-3}$ (μm)	V (cm^3)	ΔV (cm^3)	$\Delta V/\Delta P \times 10^3$ ($cm^3\ psia^{-1}$)	$D_v(r)$ ($cm^3\ \mu m^{-1}$)	$D_s(r)$ ($m^2\ \mu m^{-1}$)
50	50	2152	.085	.085	1.700	0.039	0.000
200	150	538.0	.142	.057	0.380	0.141	0.000
500	300	215.2	.251	.109	0.363	0.843	0.001
1000	500	107.6	.330	.079	0.158	1.477	0.003
5000	4000	21.52	.461	.131	0.0327	7.598	0.071
10000	5000	10.76	.633	.172	0.0344	31.970	0.594
15000	5000	7.173	.772	.139	0.0278	58.135	1.621
20000	5000	5.380	.825	.053	0.0106	39.405	1.465
30000	10000	3.587	.860	.035	0.0035	29.272	1.632
40000	10000	2.690	.862	.002	0.0002	2.974	0.221
50000	10000	2.152	.863	.001	0.0001	2.323	0.216
60000	10000	1.793	.863	.000	—	—	—

Figure 11-3 is the distribution prepared from Figure 11-2. A series of values of the ratio $\Delta V/\Delta P$ are taken from Figure 11-2 or preferably from the raw data. Each value of $\Delta V/\Delta P$ is multiplied by the pressure at the upper end of the interval and divided by the corresponding pore radius. The resulting value is plotted versus the pore radius. Table 11-1 contains the raw data from which the cumulative curve, Figure 11-2, is composed, as well as the tabularized calculations necessary to prepare the distribution curve, Figure 11-3.

Figure 11-3 is plotted semilogarithmically to include a wide range of pore radii.

4. Pore Surface Distribution

The pore surface distribution $D_s(r)$ is the surface area per unit pore radius. By chain differentials one can write

$$D_s(r) = \left(\frac{dS}{dV}\right)\left(\frac{dV}{dr}\right) \qquad [11.23]$$

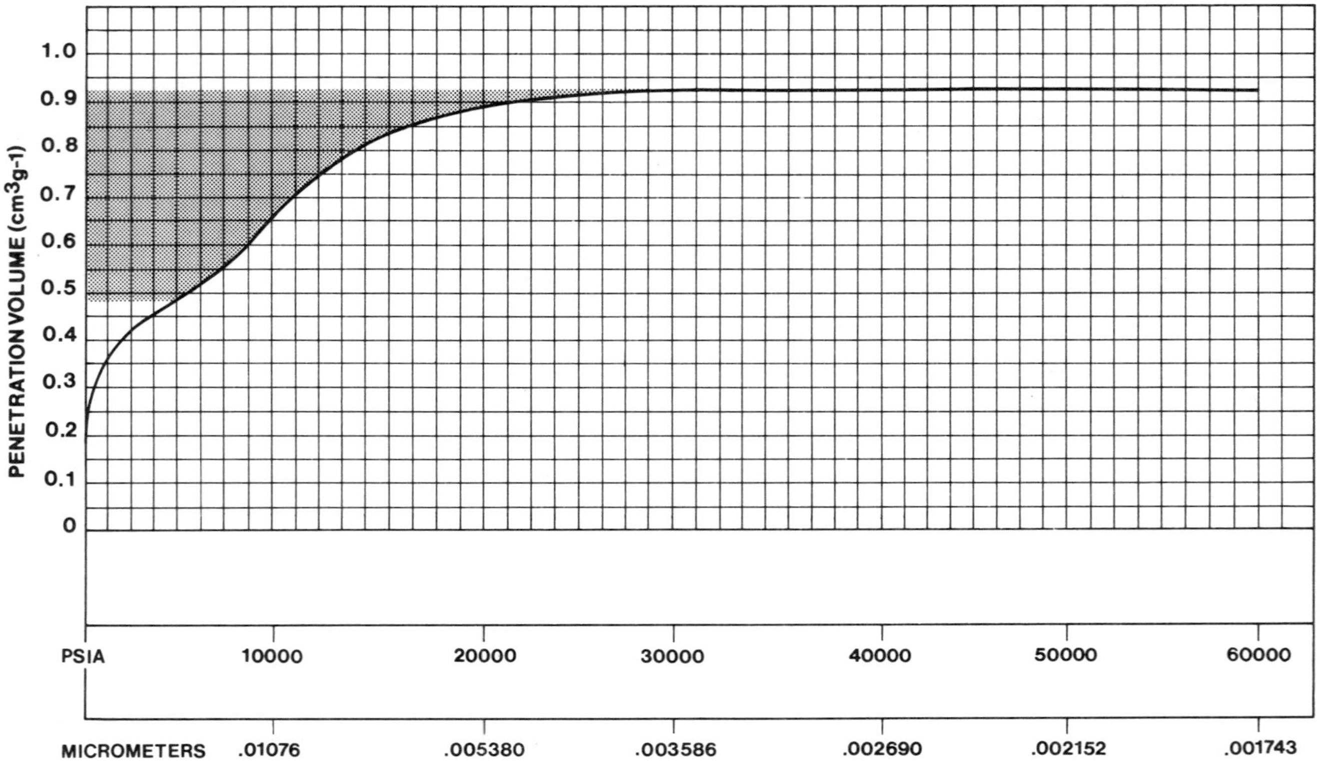

Figure 11-2 Cumulative pore volume plot.

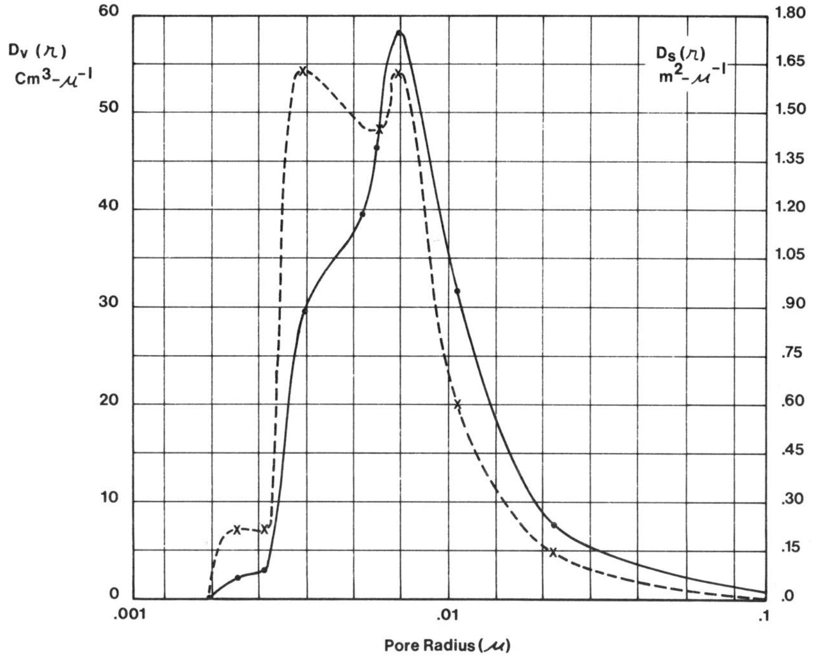

Figure 11-3 Distribution curves for volume (●), $D_v(r)$ and area (×), $D_s(r)$.

If the pores are cylinders, then

$$\frac{dS}{dV}=\frac{2}{r} \qquad [11.24]$$

and

$$D_s(r)=\frac{2}{r}\,D_v(r) \qquad [11.25]$$

The last column in Table 11-1 gives values of $D_s(r)$ which are also plotted in Figure 11-3.

The total area under the volume or area distribution curves is proportional to the total pore volume or pore area, respectively. By taking the ratio of the graphical area A_g, in any interval, to the total graph area A_t, the pore volume or surface area in an interval can be calculated by

$$V_{\text{interval}} = V_{\text{total}} \frac{A_g}{A_t} \qquad [11.26]$$

$$S_{\text{interval}} = S \frac{A_g}{A_t} \qquad [11.27]$$

5. Comparisons of Porosimetry and the Kelvin Equation

The useful range of the Kelvin equation is limited at the narrow pore end by the question of its applicability (see Chapter 9, Section 5), and at the wide pore end, measurements are limited by the rapid change of the center core radius with relative pressure.

Mercury porosimetry has somewhat the same constraints at the narrow pore end of its range, in that the same questions arise regarding the constancy of surface tension and wetting angle for mercury as exists for an adsorbate. Consequently, both methods have nearly the same lower limit which is about 20 Å for mercury intrusion, (e.g., 60,000 psia). However, at the wide pore end porosimetry does not have the limitation of the Kelvin equation and, for example, at 1.0 psia pore volumes can be measured in pores of 107 micrometer diameter or 1.07×10^4 Å.

Comparisons have been made in the range where the two methods overlap and they generally show reasonable agreement. Zwietering[85] obtained distribution curves from mercury porosimetry and the nitrogen isotherms on chromium oxide–iron oxide catalysts. Each curve showed a narrow distribution with the peak near 150 Å, the nitrogen curve being slightly narrower and higher. Using nitrogen, Joyner, Barrett, and Skold[86] obtained good agreement between the two methods. By adjusting the wetting angle for mercury on charcoals to values between 130 and 140° they were able to closely match the curve produced from the nitrogen isotherm. Cochran and Cosgrove,[87] using *n*-butane, reported total pore volumes of 0.458 $cm^3\ g^{-1}$ and 0.373 $cm^3\ g^{-1}$ with adsorption and porosimetry, respectively. The maximum pore radius was under 500 Å and they attributed the difference to *n*-butane entering pores narrower than 30 Å. Dubinin et al.[88] found that nitrogen and benzene gave type I isotherms with hysteresis on several activated carbons. The pore distributions based on the Kelvin equation agreed with porosimetry measurements.

6. Hysteresis

Generally, cumulative volume curves generated by intruding mercury (see Figure 11-2), are not followed as the pressure is lowered and mercury extrudes out of the pores. As with adsorption hysteresis, the depressurization curve lies above the pressurization curve and often the hysteresis loop does not close at all, indicating that, even when returned to zero pressure, some mercury remains within the pores. In many instances intrusion–extrusion hysteresis can be attributed to "bottleneck" shape pores. In pores of this type intrusion can not occur until sufficient pressure is attained to force mercury into the narrow neck whereupon the entire pore will fill. However, on depressurization the wide pore body will not empty until a lower pressure is reached, and if the neck is sufficiently narrow the mercury can be permanently trapped in the wide part of the pore.

Another factor contributing to hysteresis is the influence of the pore potential which can act to trap mercury once it intrudes into a pore. Also, the possibility that the contact angle changes as the mercury advances into and recedes out of a pore is still another possible cause of hysteresis.

PART TWO

EXPERIMENTAL

CHAPTER TWELVE

Adsorption Measurements —Preliminaries

1. Reference Standards

All equipment designed to measure surface area or pore volume by adsorption actually determines the quantity condensed on a solid surface at some equilibrium vapor pressure. The surface area or pore volumes and sizes are then calculated by means of an appropriate theory used to treat the adsorption data. Depending upon the apparatus employed, the adsorbed quantity is measured as volume or weight. The accuracy of adsorption apparatus is, therefore, dependent upon its ability to correctly measure either of these quantities.

If agreement is not reached between a properly calibrated apparatus and a reference standard, the reference material must be questioned. All too often materials used as standards for surface area or pore volume prove to be inadequate for a wide variety of reasons, some of which are listed below:

- Irreversible chemical reactions with atmospheric constituents, particularly water vapor and oxygen.

- Breaking and abrasion of particles incurred during handling.
- Diffusion to and from the surface, to and from the interior.
- Cold flow of some materials below but near their melting points.
- Sample separation leading to nonrepresentation.
- Variations in sample conditioning.

These and other factors must be examined before a method or apparatus is condemned for not agreeing with some other method or equipment. It is often simpler and more definitive to establish that an apparatus is properly working and calibrated than it is to use reference standards, which are subject to the hazards listed previously. Furthermore, the question must always arise regarding the merits of the technique and apparatus used to establish the standard's value in the first instance. Experimental results obtained with carefully calibrated apparatus are preferred to results that may agree with a presumed standard whose properties may change or that may have been incorrectly measured in another laboratory.

2. Other Preliminary Precautions

Other factors over which caution must be exercised when comparing measurements include gas purity and the temperature at which the data is obtained. In some instances small quantities of impurity in the adsorbate vapor can be preferentially adsorbed. On nonporous surfaces this will not severely alter the results; however, on porous samples, impurities can block the entrance to smaller pores and in extreme cases even condense in the pores, thereby drastically altering the measurement. Even when using high-purity gases a cold trap is recommended. Gases should be transferred from the pressurized tank to the apparatus with metal tubing (copper or stainless steel). Plastic tubing is not at all satisfactory because atmospheric components can often diffuse through the walls into the system. Also, many plastics contain plasticizers that can diffuse into the adsorbate as it is transferred. When metal tubing is used, it is necessary to heat the tubing to about 150°C and purge with dry nitrogen for 24 hours to remove lubricating residue used during the extruding process.

The temperature of the adsorbent and vapor with which it is in equilibrium must remain constant during the measurement. With nitrogen as the adsorbate the coolant most often used is liquid nitrogen. The temperature of liquid nitrogen is dependent on ambient pressure and the presence of impurities in the liquid which tend to elevate the boiling point. The Clapeyron–Clausius[89] equation can be used to calculate the change in liquid nitrogen temperature when the temperature is near the normal boiling point. Using 77.36°K (−195.79°C) as the normal boiling point and 1335 cal mole^{-1} as the heat of vaporization gives

$$\ln P = 15.3182 - \frac{671.867}{T} \qquad [12.1]$$

Thus, near the normal boiling point the temperature will change by 0.01°K when the equilibrium pressure changes by 0.86 mm of Hg.

3. Representative Samples

Often, for the purpose of laboratory analysis, it is necessary to obtain a small quantity of powder from a larger batch. For maximum accuracy it is necessary that the sample chosen be representative* of the larger initial quantity.

To some extent, under even slight agitation, particles tend to segregate with the finer ones settling toward the bottom of the container. When poured from a container into a conical pile the smaller particles will tend toward the center and toward the base. This behavior is caused by the small particles settling through the voids between the larger ones. In essence, the larger particles with a lower bulk density† can be thought of as floating on the smaller ones with a higher bulk density.

*Here, the term "representative" means that the sample to be analyzed must possess the same particle and pore size distributions and specific surface area as the larger quantity from which it was obtained.

†Bulk density is defined as the mass divided by the particle plus void volume.

It is generally impossible to make a segregated sample homogeneous by shaking, tumbling, or any other technique. Often these attempts only further aggravate the segregation process. Devices such as the spinning rifflers shown in Figures 12-1 and 12-2 can be used to obtain representative samples. Rifflers operate on the principle that a sample need not be homogeneous in order to be representative.

Both of the rifflers shown in Figures 12-1 and 12-2 operate by loading the powder sample into a vibrating hopper which delivers the sample down a chute into eight rotating collectors. Both the delivery and rotational rate can be controlled.

The sample, when loaded in the hopper, will be segregated. Therefore, at any depth l there will exist a particle diameter gradient $\Delta D/\Delta l$. The powder settles as it is delivered to the collectors

Figure 12-1 Sieving riffler. Courtesy of Quantachrome Corporation.

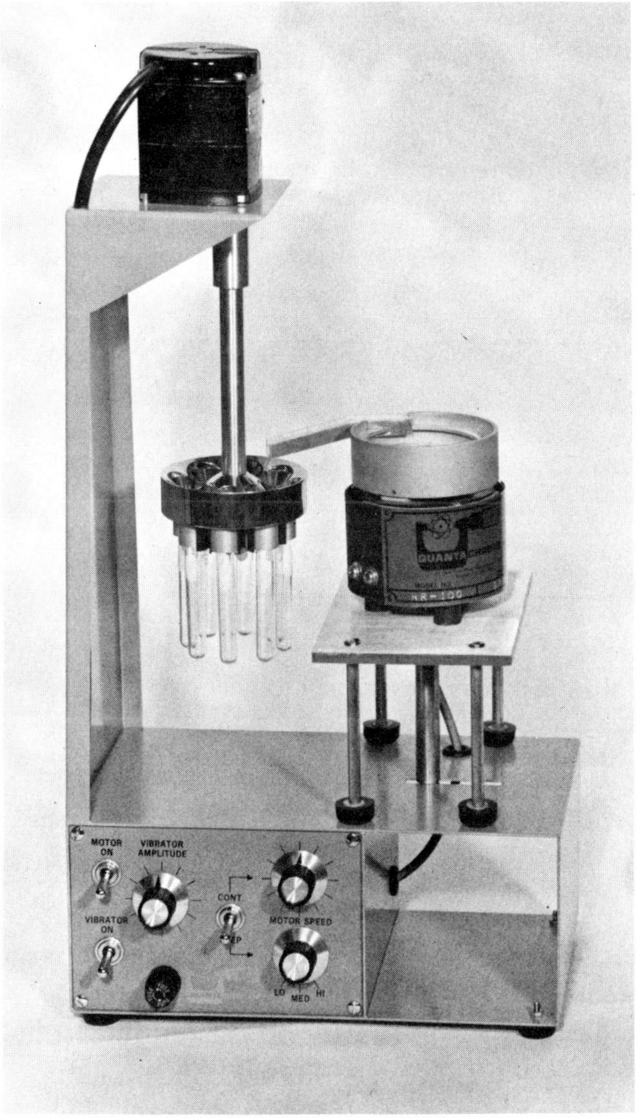

Figure 12-2 Micro riffler. Courtesy of Quantachrome Corporation.

at the rate $\Delta l/\Delta t$. Then,

$$\frac{\Delta D}{\Delta l}\cdot\frac{\Delta l}{\Delta t}=\frac{\Delta D}{\Delta t} \quad [12.2]$$

which is the time rate of change of particle diameters leaving the hopper. If ψ is defined as the change in particle diameter entering each collector per revolution, then

$$\psi=\frac{1}{8}\frac{\Delta D}{\Delta t}\tau \quad [12.3]$$

where τ is the time per revolution. Substituting equation [12.2] into [12.3] gives

$$\psi=\frac{1}{8}\frac{\Delta D}{\Delta l}\cdot\frac{\Delta l}{\Delta t}\tau \quad [12.4]$$

Recognizing that $\Delta l/\Delta t$ is proportional to the feed rate F yields

$$\psi=K\frac{\Delta D}{\Delta l}F\tau \quad [12.5]$$

Equation [12.5] asserts that the change in particle diameter entering each collector can be made as small as necessary by decreasing the feed rate or increasing the collector's rotational rate. However, Hatton[90] argues that it is preferable to slow the feed rate rather than increase the rotational rate.

When the entire sample has been delivered, each collector will contain powder exactly representative of the initial batch. Each collector will also contain a size gradient from top to bottom. Therefore, if the quantity required for analysis is less than the amount in any single collector, the process must be repeated. This is achieved by placing the contents of one collector back into the hopper and reriffling. The Micro Riffler, Figure 12-2, can riffle samples less than one gram to accommodate the need for small final samples.

The large riffler, Figure 12-1, has an optional sieve that can be placed on top of the hopper to exclude particles above a required size. The advantage of this arrangement lies in the fact that a single particle of 100 μm radius has the same weight as one million one-micrometer particles with only one-hundredth the surface area.

If only a few of the large particles are present they may not be properly represented in the final sample.

Often it is thought that the effectiveness of a riffler can be demonstrated by the uniformity of weight accumulated in each collector. This reasoning is incorrect if one considers that each collector will necessarily acquire a slightly different amount of sample if the diameters vary slightly. The only correct test for the effective performance of a riffler is to compare the contents of each collector in terms of particle size distribution or specific surface area.

Experiments with silica powder chosen from five depths in a 2-pound container gave surface areas from top to bottom of 9.8, 10.2, 10.4, 10.5, and 10.7 $m^2\ g^{-1}$. When the same sample was poured into a conical pile, five random samples produced surface areas of 10.3, 11.0, 10.4, 10.0 and 10.6 $m^2\ g^{-1}$. However, when the sample was riffled in a spinning riffler with three size reductions, the subsequent analysis of the contents of five collectors gave 10.2, 10.1, 10.2, 10.2, and 10.1 $m^2\ g^{-1}$ as the specific surface area.

4. Sample Conditioning

Before a sample can be analyzed it is necessary to remove contaminating materials which can alter the surface potential and block or fill pores. Sample conditioning can be accomplished by vacuum pumping or purging with an inert gas. Both methods require the use of elevated temperatures to hasten the rate at which contaminants leave the surface. Caution must be used not to reach temperatures at which the sample's surface properties can be altered. Melting, dehydration, sintering, and decomposition are processes that can drastically alter surface properties. A test that can be used to establish that excessive outgassing temperatures have not been used is to determine the surface area of a duplicate sample, after outgassing at higher temperature, for a longer time. Duplication of results indicates the initial outgassing conditions were satisfactory unless both analysis were performed on a completely degraded sample.

When a vacuum is used to outgas, precautions must be taken to prevent the initial surge from blowing powder throughout the system. In a well-sealed system outgassing can be taken as completed when no pressure increase can be detected after isolation from the pump. For routine work generally 10^{-3}–10^{-5} torr (mm of Hg) is sufficient. The usual precautions such as a baffle and a cold trap should be taken to insure no contamination by the pump oil.

When outgassing samples by purging, the purge gas must be of high purity and passed through a cold trap to remove traces of organic compounds or water vapor. Completion of outgassing can be determined by passing the effluent over a thermal conductivity detector which can detect impurities in the effluent as low as a few parts per million. Figure 12-3 illustrates a convenient flow system for outgassing using a purge. Detector D_2 compares the effluent from the sample cell with clean gas entering detector D_1. A reference or zero impurity signal is obtained by switching valve *e* to bypass the sample cell. Helium or hydrogen will serve best as the purge gas, (see Chapter 14, Section 3).

In those instances where samples cannot be heated the method of repetitive cycling, investigated by Lopez-Gonzales, Carpenter, and Deitz[91] can be utilized. They found that by repetitive adsorption and desorption the surface can be adequately cleaned to allow

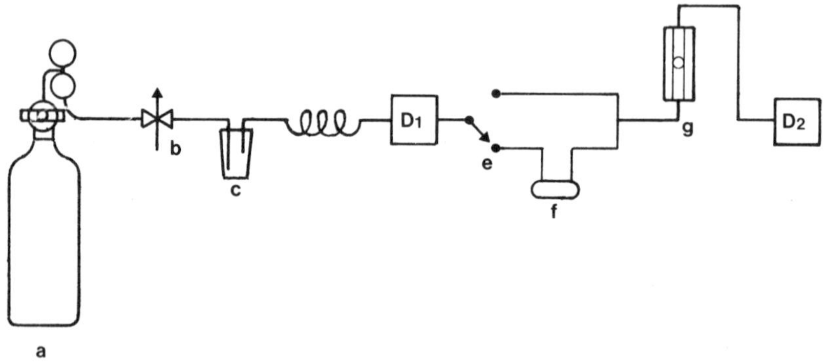

Figure 12-3a Flow outgassing with thermal conductivity detector. *a*, purge gas source; *b*, needle valve; *c*, cold tap; D_1 and D_2 thermal conductivity detectors; *e*, flow selector valve; *f*, cell with powder; *g*, flow meter.

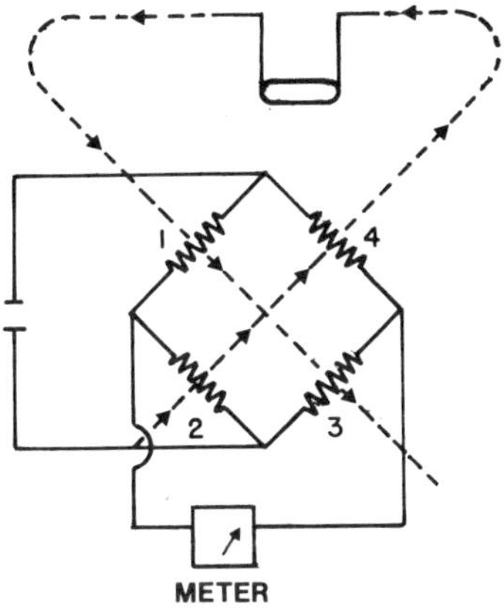

Figure 12-3b Heated filaments 1 and 3 are D_2, 2 and 4 are D_1. Dotted line is gas flow through sample cell. Signal meter is shown at bottom.

reproducible measurements. Usually three to six cycles are sufficient to produce a well-behaved surface. The flow diagram of Figure 12-3 is well suited to this type of procedure.

CHAPTER THIRTEEN

Vacuum Volumetric Measurements

1. Nitrogen Adsorption

Many types of vacuum adsorption apparatus have been developed[92–95] and no doubt every laboratory where serious adsorption measurements are made has equipment with certain unique features. The number of variations are limited only by the need and ingenuity of the users. However, all vacuum adsorption systems have certain essential features, including, a vacuum pump, two gas supplies, a sample container, calibrated volume, manometer, and a coolant.

Figure 13-1 shows a simplified vacuum adsorption apparatus that is suitable for nitrogen adsorption.

The volumes V_a, V_b, and V_c, between fiducial marks, are carefully calibrated prior to sealing in place. Calibration is achieved by filling with mercury, emptying, and weighing. The manifold volume V_m is determined by admitting helium through stopcock 2, into the evacuated manifold. Before and after admitting helium, the mercury in the left arm of the manometer is brought to the fiducial mark by adjusting stopcock 8. With helium in the manifold the pressure P

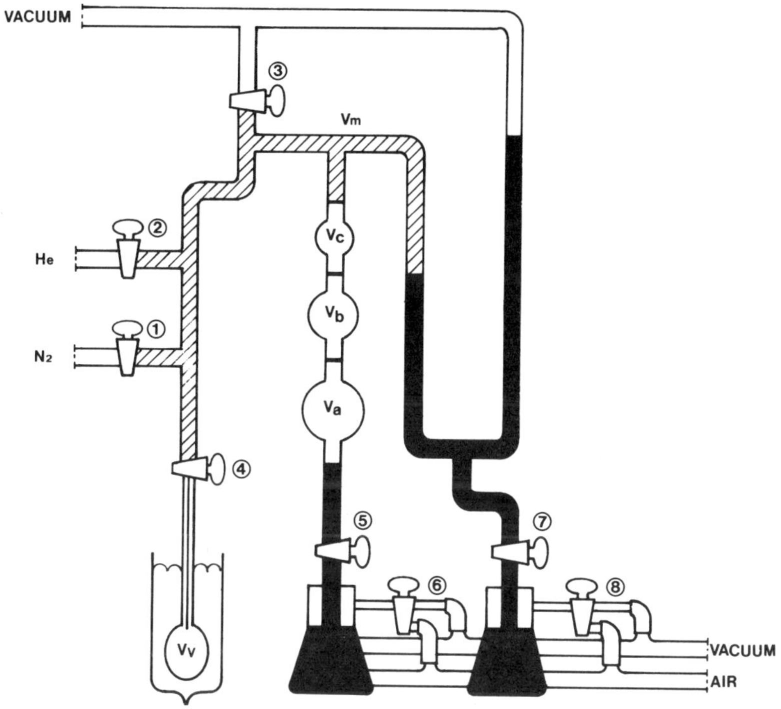

Figure 13-1 Simplified vacuum adsorption apparatus. Shaded areas represent mercury. Heavy horizontal lines are fiducial marks. Striped volume is manifold volume V_m.

is read as the difference in mercury levels in the manometer arms. When stopcock 5 is turned, mercury is allowed to fill the lower bulb V_a by rising from the lowest fiducial mark to the one above. The manometer is again brought to its fiducial mark and pressure P' is noted. From the ideal gas equation one obtains

$$P(V_m + V_a + V_b + V_c) = P'(V_m + V_b + V_c) \qquad [13.1]$$

then

$$V_m = \frac{PV_a - (P' - P)(V_b + V_c)}{(P' - P)} \qquad [13.2]$$

The sample cell is not completely filled by the adsorbent; it contains a void volume V_v which is the volume not occupied by adsorbent, up to stopcock 4. If the calibrated volumes and manifold are filled with nitrogen at pressure P_1 and stopcock 4 is opened the pressure will fall because of expansion and adsorption of nitrogen in the evacuated cell. After equilibrium is attained and the mercury in the manometer is returned to the fiducial mark the new pressure P_1 is noted. The number of moles adsorbed is given by

$$n_1=\frac{P_1(V_a+V_b+V_c+V_m)}{RT}-\frac{P_1'(V_a+V_b+V_c+V_m)}{RT}+\frac{P_1'V_v}{RT_v} \qquad [13.3]$$

where T is the temperature of the calibrated volumes and manifold and T_v is the temperature within the sample cell, assumed to be uniform. The value of V_v can be determined by using helium gas and repeating the preceding steps, the so-called blank run, during which helium is not significantly adsorbed at liquid nitrogen temperature. Since no adsorption occurs, n_1 in equation [13.3] is zero and V_v can be calculated.

Expressing pressure in atmospheres and all volumes in cubic centimeters, equation [13.3] can be rewritten in terms of the volume adsorbed V_1 at standard temperature (273.2°K) and pressure (1.0 atm)

$$V_1=P_1V_0^{\ddagger}-P_1'(V_0^{\ddagger}-V_v^{\ddagger}) \qquad [13.4]$$

where V_0 is the sum of the calibration volume $V_a+V_b+V_c$ and the manifold volume V_m. To reduce all gas volumes to standard conditions the following relations are used:

$$V_0^{\ddagger}=\frac{V_0 273.2}{760T} \qquad [13.5]$$

$$V_v^{\ddagger}=V_v\frac{273.2}{760T_v} \qquad [13.6]$$

For subsequent data points stopcock 4 is closed and additional nitrogen is admitted into the calibrated volumes and manifold. When stopcock 4 is opened the pressure will fall from P_2 to P_2' and the volume adsorbed under standard conditions is given by

$$V_2=[P_1V_0^{\ddagger}+(P_2-P_1')V_0^{\ddagger}]-P_2'(V_0-V_v^{\ddagger}) \qquad [13.7]$$

In general, for the nth step the volume adsorbed is written as

$$V_n=[P_1V_0^{\ddagger}+(P_2+P_1')V_0^{\ddagger}+\cdots+(P_n-P_{n-1}')V_0^{\ddagger}]-P_n'(V_0^{\ddagger}-V_v^{\ddagger}) \qquad [13.8]$$

The terms in the bracket represent the sum of the gas added to the system, and the second term contains the equilibrium quantity of gas remaining unadsorbed.

The purpose of several calibrated volumes V_a, V_b, and V_c, is to enable more than one data point to be obtained from each filling. This is accomplished by raising the mercury level to arrive at a new initial pressure before opening stopcock 4. Also, the ability to adjust the calibrated volume gives some control over the pressure difference when small or large amounts of gas are adsorbed.

Points on the desorption isotherm are obtained by first adsorbing at some relative pressure close to unity. Then, stopcock 4 is closed and the calibrated volumes and manifold are evacuated. Stopcock 4 is opened which permits desorption to occur. The equilibrium pressure is obtained from the manometer and the volumes desorbed are calculated using equation [13.8] except that P' is now the pressure after desorption.

Attempts to measure points on the desorption isotherm without first adsorbing near saturation will result in scanning the hysteresis loop from the adsorption to the desorption curve, (see Figure 14-13 and Chapter 14, Section 6).

The errors in the vacuum volumetric method arise principly from two sources. The error in measured doses is cumulative requiring very accurate calibration. Also, the void volume correction becomes more significant at higher pressures when measuring low surface areas because more molecules are retained in the void volume than are adsorbed.

2. Deviations From Ideality

The gas contained in the void volume of the sample cell must be corrected for deviations from ideality. Prior to reducing the void volume V_v to standard conditions using equation [13.6] the measured volume should be corrected for nonideality. For nitrogen, Emmett and Brunauer[96] derived the appropriate correction, which is linear with pressure, from the Van der Waals equation. Using V_v' as the corrected void volume, their correction is

$$V_v' = V_v\left(1 + \frac{\alpha P}{760}\right) \qquad [13.9]$$

where $\alpha = 6.6 \times 10^{-5}$ for nitrogen and P is in torr. Table 13-1 gives α values for several gases.

Table 13-1. Correction Factors for Nonideality[a]

Gas	Temperature (°C)	α
N_2	−195.8	6.58×10^{-5}
	−183	3.78×10^{-5}
O_2	−183	4.17×10^{-5}
A	−195.8	11.4×10^{-5}
	−183	3.94×10^{-5}
CO	−183	3.42×10^{-5}
CH_4	−140	7.79×10^{-5}
NO	−140	5.26×10^{-5}
N_2O	−78	7.68×10^{-5}
CO_2	−78	2.75×10^{-5}
	25	1.559×10^{-5}
n-C_4H_{10}	0	14.2×10^{-5}
	25	4.21×10^{-5}

[a]Calculated from Emmett and Brunauer.[96]

3. Sample Cells

Sample cells should be designed to minimize the void volume. Samples with adequately low surface area may adsorb less volume than is required to fill the void volume. In such instances, large errors can be generated unless the void volume is accurately measured and reduced to as small a value as possible.

Ideally, the sample should be placed in an open-ended glass cell and then closed with a torch prior to sealing onto the adsorption apparatus.

The stem of the sample cell should be capillary tubing to further reduce the void volume and, equally important, to minimize the volume of the cell stem that emerges from the coolant bath that is not at a uniform temperature.

4. Evacuation and Outgassing

After attaching the sample and sample cell to the apparatus, caution must be used to prevent powder from blowing out during the initial evacuation of the sample cell. If this occurs it is necessary to recalibrate the various parts of the apparatus. A plug of glass wool inserted at the junction of the capillary tube and the wider part of the sample cell will prevent particles from entering the manifold. Powder trapped in the glass wool, however, will contribute to adsorption and represents a potential source of error.

During outgassing, when the sample is being heated, a vacuum must be maintained in the rest of the apparatus to prevent condensation of vapor in the cooler parts of the equipment.

5. Temperature Control

For maximum accuracy the calibrated volumes and the manifold should be maintained at constant temperature by using a water or constant temperature air jacket. The liquid nitrogen level should be maintained as near constant as possible to prevent any variation in the effective volume of the sample cell.

6. Isotherms

Because the shape of the isotherm establishes the quantity adsorbed at various relative pressures, it is often difficult to predict a priori the exact dosing quantity to admit to the calibrated volumes in order to obtain the desired equilibrium pressure. If the first dose of adsorbate equilibrates at a relative pressure higher than desired because of insufficient adsorption, it is best to reevacuate the sample and admit a smaller initial dose. Thereafter, smaller dose requirements can be anticipated for a subsequent data points. Conversely, if the first dose equilibrates at a lower relative pressure than desired, it is necessary only to dose a second time to attain a higher equilibrium pressure.

Because of the dosing technique associated with the vacuum volumetric method there exists a potential source of error which in principle cannot be avoided on samples that are slow to equilibrate. Figure 13-2 represents a plot of the pressure in the sample cell versus time. The pressure up to time t_1 represents the equilibrium pressure in the sample cell before a new quantity of adsorbate is admitted. At t_1 adsorbate is admitted into the sample cell accompanied by a rapid pressure rise. The pressure then gradually decreases to a new equilibrium value at t_2. If the decay to the new equilibrium pressure is slow, it is possible for open or accessible parts of the surface, such as the interior of wide pores, to contain more adsorbate before equilibrium is attained than after equilibrium is established. Less accessible regions, such as the interior of long narrow pores, will adsorb slowly, and as the pressure falls because of adsorption in these pores, desorption must occur from the more accessible pores which tend to equilibrate more rapidly. If a porous sample is subjected to increasing pressure followed by decreasing pressure an unusual situation develops. This is because desorption cannot occur along the adsorption isotherm and if the

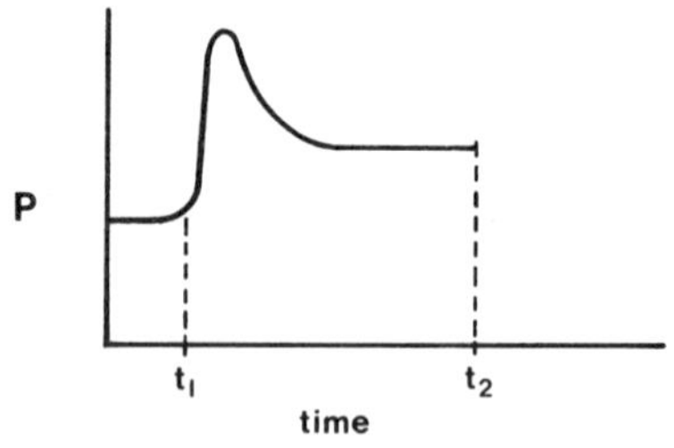

Figure 13-2 Adsorbate pressure before and after dosing, showing pressure overshoot before equilibrium is attained.

highest pressure reached is less than the saturated pressure, the quantity adsorbed will slide off the adsorption isotherm toward the desorption isotherm. Thus, depending upon the final equilibrium pressure, the data point will lie somewhere between the two curves that comprise the hysteresis loop.

It may well be that the effect described previously occurs to some extent on any sample that contains pores with one or more constrictions. If the constriction is only slightly wider than the adsorbate dimensions, entrance into the wide inner pore will be hindered because of the strong interaction with the walls in the narrow portion. Adsorbate molecules will be able to pass through the constriction only if they possess sufficient energy to overcome the higher potential within this region. This situation is similar to the Arrhenius[97] concept of an activation energy used in chemical kinetics. According to Arrhenius the specific reaction rate k is given by

$$k = Ae^{-E/RT} \qquad [13.10]$$

where E is the energy molecules must possess in order to react and A is related to the collision frequency with which molecules collide. By analogy, the rate at which molecules can penetrate through a constriction would be given by

$$\text{Rate} = Ae^{-E_{ad}/RT} \qquad [13.11]$$

Here, E_{ad} is the adsorption potential within the constriction. Equation [13.11] predicts that adsorption in pores with one or more constrictions will proceed more rapidly at elevated temperatures. Several examples of this type of "activated physical adsorption" have been found.[98-101]

If the adsorption potential in a constriction is 2 kcal more than in a wide pore, the ratio of adsorption rates at liquid nitrogen temperature, assuming the same preexponential value, will be approximately

$$\frac{\text{Rate}_1}{\text{Rate}_2} = \frac{A_1}{A_2} e^{2000/2 \times 77} = 2.3 \times 10^6 \qquad [13.12]$$

Therefore, it may take weeks or months for equilibrium to be established in the constricted pores. A rate ratio of only 2 or 3 would be sufficient to cause adsorption and then desorption to occur from the accessible or wider pores leading to the data point residing in the hysteresis regions.

The consequences of this type of "activated physical adsorption" is not only that the quantity adsorbed can lie off the isotherm but also that the measured quantity of adsorption is far less than the equilibrium value. No experiments have been conducted to illustrate whether the quantity adsorbed lies within the hysteresis loop. The occasional failure of the vacuum volumetric method to agree with the dynamic method, which is not subject to any pressure overshoot, may in part be attributed to this phenomenon.

7. Low Surface Areas

By using small calibrated volumes, carefully measuring the void volume and maintaining constant temperature by thermostating the adsorption apparatus, surface areas as low as approximately 1.0 m^2 can be measured. The principal obstacle to measuring areas lower than this with nitrogen and a volumetric system lies in accurately measuring the void volume. On small areas, the quantity of adsorbate remaining in the void volume is large compared to the amount adsorbed and, indeed, the void volume error can be larger than the volume adsorbed.

The number of molecules trapped in the void volume can be reduced by using adsorbates with low vapor pressures. Beebe, Beckwith, and Honig[102] were the first to use krypton for this purpose. Litvan[103] reports the vapor pressure of krypton at liquid nitrogen temperature as 2.63 torrs. Therefore, the amount of krypton remaining in the void volume, when monolayer coverage occurs, will be much less than nitrogen, whereas the amount of adsorption will be only slightly less by approximately the ratio of cross-sectional areas or about 16.2/19.5.

To measure the low pressures associated with krypton adsorption a thermocouple or thermistor can be used after calibration against a McLeod gauge.

Argon at liquid nitrogen temperature exhibits an equilibrium pressure of 187 torrs. It offers the advantage of lower vapor pressure than nitrogen, which will reduce the void volume error while retaining ease of pressure measurements. However the cross-sectional area of argon is not well established and appears to vary according to the surface on which it is adsorbed.

When using adsorbents with low vapor pressure no correction for nonideality is necessary. However, another type of correction is required due to "thermal transpiration"[104] which occurs if two parts of a system contain the same gas at the same pressure but at different temperatures. Thermal transpiration can be understood by considering the number of collisions n per square centimeter per second which a gas makes with the walls of its container

$$n = \frac{\bar{N}P}{(2\pi\bar{M}RT)^{1/2}} \qquad [13.13]$$

A tube connecting the two parts of a system will admit more molecules on the cold side than on the warm side. Consequently, some gas will flow from the cold to the warm ends of the tube until the pressure on the warm side increases sufficiently to permit the following equality;

$$n_{\text{cold}} = n_{\text{warm}} \qquad [13.14]$$

so that

$$\frac{P_{\text{cold}}}{T_{\text{cold}}^{1/2}} = \frac{P_{\text{warm}}}{T_{\text{warm}}^{1/2}} \qquad [13.15]$$

then

$$P_{\text{warm}} = \left(\frac{T_{\text{warm}}}{T_{\text{cold}}}\right)^{1/2} P_{\text{cold}} \qquad [13.16]$$

For the effects of thermal transpiration to be observed it is necessary that the mean free molecular path be greater than the tube diameter. At higher pressures and smaller mean free paths, molecular collisions destroy the effect.

In the pressure range useful for krypton adsorption the mean free path is approximately equal to the diameter of the stem of the sample cell. Equation [13.16] is not applicable in this range and an empirical equation devised by Rosenberg[105] must be used. His equation is

$$P_c = P_w \left[1 - \frac{0.49}{37.2(dP_w)^2 + 14.45dP_w + 1} \right] \qquad [13.17]$$

where P_c and P_w refer to the pressure (torr) in the sample cell at 77°K and room temperature, respectively. d is the inside diameter of the sample cell stem (mm).

8. Saturated Vapor Pressure, P_0 of Nitrogen

The term P_0 is defined as the saturated equilibrium vapor pressure exhibited by pure nitrogen contained in the sample cell when immersed in liquid nitrogen coolant. The temperature of the liquid nitrogen coolant, usually held in a Dewar flask, is always found to be somewhat different than the normal boiling point because of dissolved impurities from the atmosphere and because of ambient pressure fluctuations. Dissolved impurities usually increase the bath temperature sufficiently to cause the vapor pressure of pure liquid nitrogen within the sample cell to increase by 10–20 torrs above ambient pressure. Therefore, adding 15 torrs to ambient pressure will usually give the correct saturated pressure to within 5 torrs.

When measuring surface areas, the effect of using an incorrect saturated pressure is further reduced by the nature of the BET plot since both the ordinate and abscissa will deviate in the same direction leaving the slope nearly constant. For BET C values near 100, the slope = 1000 and the intercept = 10; an error of 15 torrs in a total 760 torrs or about 2%, will produce less than 1% error in surface area.

When pore sizes are analyzed above the usual BET range of relative pressures, the error associated with saturated pressure measurements is moot when compared to the many other assumptions.

When highly accurate P_0 measurements are necessary, a platinum resistance thermometer can be placed in the bath adjacent to the sample cell. Equation [12.1] can then be used to calculate the nitrogen vapor pressure. Alternatively and perhaps preferably, the vapor pressure can be measured directly by condensing nitrogen in a cell contained in the coolant and connected directly to a manometer or sensitive pressure gauge.

CHAPTER FOURTEEN

Dynamic Methods

1. Influence of Helium

In 1951, Loebenstein and Deitz[106] described an innovative technique that did not require the use of a vacuum. They adsorbed nitrogen from a mixture of nitrogen and helium which was passed back and forth over the sample between two burettes by raising and lowering attached mercury columns. Equilibrium was established by noting no further change in pressure with additional cycles. The quantity adsorbed was determined by the pressure decrease at constant volume. Successive data points are acquired by adding more nitrogen to the system.

The results obtained by Loebenstein and Deitz agreed with vacuum volumetric measurements on a large variety of samples with a wide range of surface areas. They were able also to establish that the quantities of nitrogen adsorbed were independent of the presence of helium.

Because all the methods mentioned in this chapter require that the adsorbate be mixed with an inert carrier gas, usually helium, some discussion regarding the possible influence of helium is required.

The forces leading to liquefaction and adsorption are the same in origin and magnitude. Gases substantially above their critical temperature T_c cannot be liquefied because their thermal energy is sufficient to overcome their intermolecular potential. Although the adsorption potential of a gas can be greater than the intermolecular potential, helium is, nevertheless, not adsorbed at liquid nitrogen temperature (77° K) because this temperature is still more than 14 times helium's critical temperature (5.3° K).

Furthermore, in order to be considered adsorbed a molecule must reside on the surface for a time τ at least as long as one vibrational cycle of the adsorbent normal to the surface. The time for one vibration is usually of the order 10^{-13} seconds which by equation [14.1] makes τ about 2×10^{-13} seconds at 77° K, (see equation [5.4] and the following paragraph).

$$\tau = 10^{-13} C^{100/RT} = 1.91 \times 10^{-13} \text{ sec} \qquad [14.1]$$

The value of 100 cal mole^{-1} chosen for the adsorption energy of helium is consistent with the fact that helium has no dipole or quadrapole and is only slightly polarizable. Thus, it will minimally interact with any surface. Based upon reflections of a helium beam from LiF and NaCl cleaved surfaces, de Boer[107] estimates the adsorption energy to be less than 100 cal mole^{-1}.

At 77° K the velocity of a helium atom is 638 m sec^{-1} so that in 1.91×10^{-13} sec it will travel $1.91 \times 10^{-13} \times 638 \times 10^{10}$ or 1.2 Å. Thus, the condition that the adsorbate reside near the surface for one vibrational cycle is fulfilled by the normal velocity of helium and not by virtue of being adsorbed. Stated in alternate terms, the density of helium near a solid surface at 77° K is independent of the surface and is the same as the density remote from the surface.

Molecular collisions with the adsorbed film by helium will certainly be no more destructive than collisions made by the adsorbate. In fact, helium collisions will be less disruptive of the adsorbed film structure since helium's velocity is, on the average, 2.6 times greater than that of nitrogen at the same temperature, while a nitrogen molecule is 7 times heavier. Thus, the momentum exchange due to nitrogen collisions will be the more disruptive.

The thermal energy of helium at 77° K is about 220 cal mole^{-1}. The heat of vaporization of nitrogen at 77° K is 1335 cal mole^{-1}, which may be taken as the minimum heat of adsorption. A complete exchange of thermal energy during collisions between a helium atom and an adsorbed nitrogen molecule would not be sufficient to cause desorption of the nitrogen.

A further point to be made is that the extent of helium's influence, if any, as a carrier gas in a dynamic system must be weighed against its effect when used to calibrate the void volume in a vacuum system.

2. Nelson and Eggertsen[108] Continuous Flow Method

Nelson and Eggertsen, in 1958, extended the Loebenstein and Dietz technique by continuously flowing a mixture of helium and nitrogen through the powder bed. They used a hot wire thermal conductivity detector to sense the change in effluent gas composition during adsorption and desorption, when the sample cell is immersed into and removed from the bath, respectively.

Figure 14-1 illustrates a simplified continuous flow apparatus. Figure 14-2 is a schematic of the flow path arrangement using a four-filament thermal conductivity bridge.

In Figure 14-1, a mixture of adsorbent and carrier gas of known concentration is admitted into the apparatus at *a*. Valve V_1 is used to control the flow rate. When the system has been purged, the detectors are "zeroed" by balancing the bridge, (see Figure 14-8).

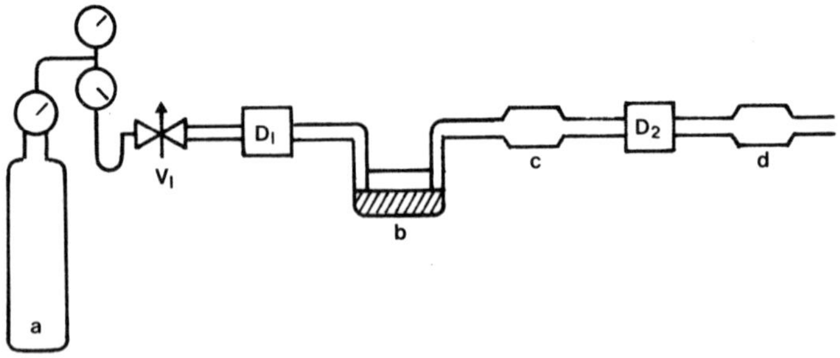

Figure 14-1 Simplified continuous flow apparatus.

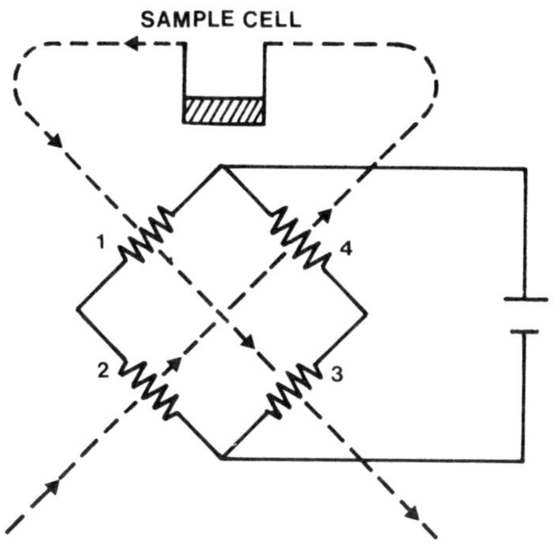

Figure 14-2 Flow path using a four-filament bridge.

As the sample cell *b* is immersed in the coolant, adsorption commences and detector D_2 senses the increased helium concentration. Upon completion of adsorption, D_2 again detects the same concentration as D_1 and the signal returns to zero. When the coolant is removed, desorption occurs as the sample warms and detector D_2 senses the increased nitrogen concentration. Upon completion of desorption the detectors again sense the same concentration and the signal returns to its initial zero value. Volume *c* is a wide segment of tubing used to decrease the linear flow velocity of the gas to insure its return to ambient temperature prior to entering D_2. Volume *d* is also a wide segment of tubing which acts as a ballast in order to prevent air from being sucked into D_2 as the cell is cooled and the gas contracts. Figure 14-3 illustrates the detector signals due to adsorption and subsequent desorption.

Figures 14-4 and 14-5 illustrate a parallel flow arrangement which has the advantage of requiring shorter purge times when changing gas composition but is somewhat more wasteful of the mixed gases. The symbols shown in Figure 14-4 have the same meaning as they do in Figure 14-1.

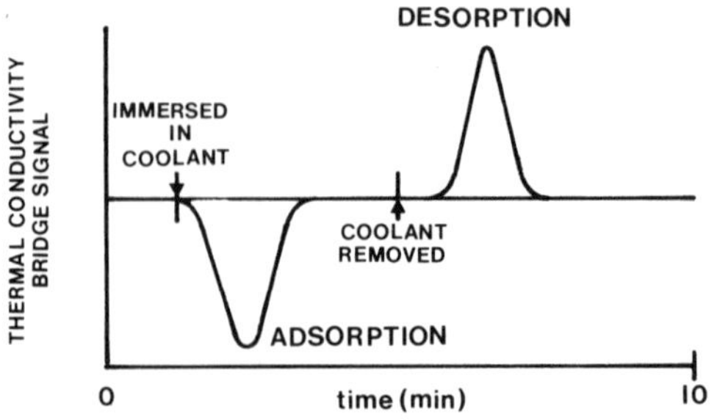

Figure 14-3 Adsorption and desorption peaks in a continuous flow apparatus.

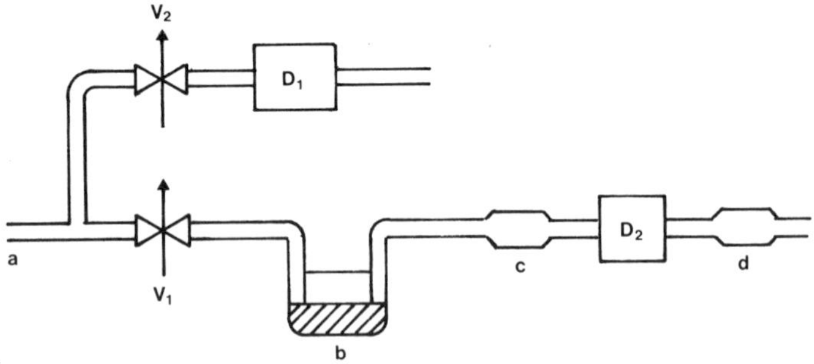

Figure 14-4 Parallel flow circuit.

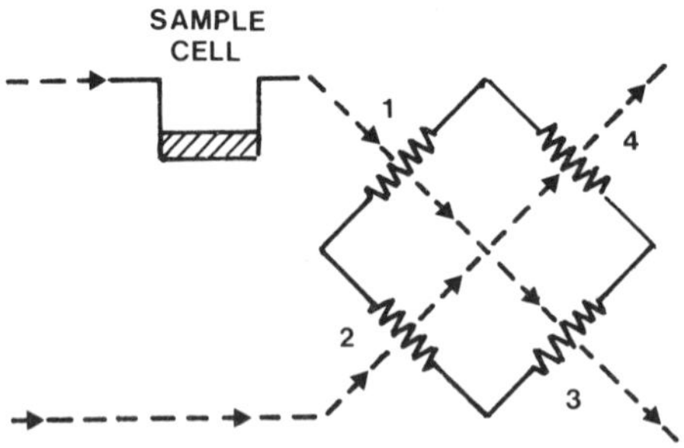

Figure 14-5 Parallel flow path using a four filament bridge. Filaments 1 and 3 are D_2, 2 and 4 are D_1, dotted lines are gas flow paths.

3. Carrier Gas and Detector Sensitivity

To be effective, the carrier gas must fulfill two requirements. First, it cannot be adsorbed at the coolant temperature; second, it must possess a thermal conductivity sufficiently different from that of the adsorbate that small concentration changes can be detected.

To understand the effect of the carrier gas on the response of the thermal conductivity detector, consider the steady state condition that prevails when the resistive heat generated in the hot wire filament is exactly balanced by the heat conducted away by the gas. This condition is described by equation [14.2]

$$i^2R = ck(t_f - t_w) \qquad [14.2]$$

where i is the filament current, R is the filament resistance, k is the thermal conductivity of the gas mixture, and t_f and t_w are the filament and the wall temperatures, respectively. The constant c is a cell constant that reflects the cell geometry and the separation of the filament from the wall which acts as the heat sink.

When the gas composition is altered due to adsorption or desorption, the value of k changes by Δk which in turn alters the filaments temperature by Δt. Under the new conditions equation [14.2] can be rewritten as

$$i^2R = c(k + \Delta k)(t_f + \Delta t_f - t_w) \qquad [14.3]$$

Equating the right-hand side of equations [14.2] and [14.3] gives

$$k(t_f - t_w) = (k + \Delta k)(t_f + \Delta t_f - t_w) \qquad [14.4]$$

By neglecting the term $\Delta k \Delta t_f$, a second-order effect, equation [14.4] rearranges to

$$\Delta t_f = \frac{\Delta k}{k}(t_w - t_f) \qquad [14.5]$$

The change in filament resistance ΔR is directly proportional to the small temperature change Δt_f and is given by

$$\Delta R = \alpha R \Delta t_f \qquad [14.6]$$

where α is the filament's temperature coefficient, dependent on the composition of the filament, and R is the filament resistance at temperature t_f; thus

$$\Delta R = \alpha R \frac{\Delta k}{k} (t_w - t_f) \qquad [14.7]$$

Equation [14.7] requires that Δk be as large as possible for maximum response under a fixed set of operating conditions.

A fortunate set of circumstances leads to a situation in which the same molecular properties that impart minimal interactions or adsorption potentials also lead to the highest thermal conductivities. Molecules of large mass and many degrees of vibrational and rotational freedom tend to be more polarizable and possess dipoles and quadrapoles which give them higher boiling points and stronger interactions with surfaces. These same properties tend to reduce their effectiveness as thermal conductors.

Helium possesses only three degrees of translation freedom and hydrogen the same, plus two rotational and one vibrational degree. However, because of hydrogen's light weight, it has the highest thermal conductivity of all gases, followed by helium. Either of these two gases fulfills the requirement for adequately high thermal conductivities so that Δk in equation [14.7] will be sufficiently large to give good sensitivity with any adsorbate. Helium, however, is usually used in continuous flow analysis because of the hazards associated with hydrogen.

Figure 14-6 is a plot of the thermal conductivity of mixtures of helium and nitrogen obtained on an apparatus similar to that described in the next section. Characteristically, the thermal conductivity of most mixtures does not vary linearly with concentration. The slope of the curve at any point determines the value of Δk and, therefore, the detector response. Figure 14-6 also illustrates that the greater the difference between thermal conductivities of the adsorbate and carrier gas, the higher will be the slope and, therefore, the detector response.

The shape of the curve shown in Figure 14-6 is fortuitous in as far as the continuous flow method is concerned. For reasons to be discussed later, the desorption signal (see Figure 14-3) is generally used to calculate the volume adsorbed. When, for example, 1.0 cm^3

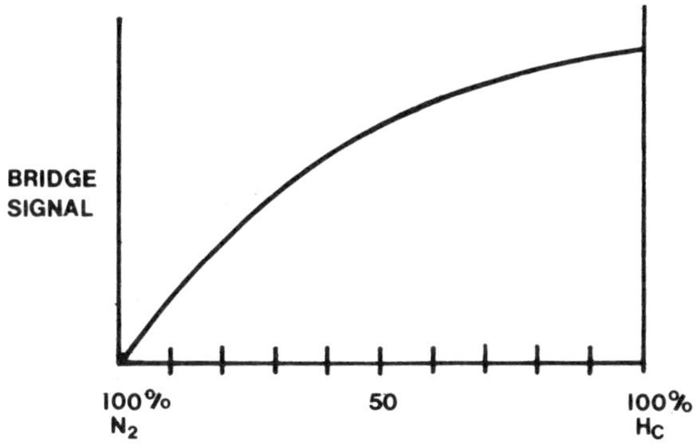

Figure 14-6 Thermal conductivity bridge unbalance. He through D_1 and He–N_2 mixture through D_2. He thermal conductivity, 0°C = 33.6 $\times 10^{-5}$ cal sec^{-1} cm^{-2}°C^{-1} cm^{-1}. N_2 thermal conductivity, 0°C = 5.68 $\times 10^{-5}$ cal sec^{-1} cm^{-2}°C^{-1} cm^{-1}. From *Chemistry Handbook*, 8th ed., 1952.

of nitrogen is desorbed into 9.0 cm^3 of helium, the concentration change is 10%. However, when 1.0 cm^3 of nitrogen is desorbed into 9.0 cm^3 of a 90% nitrogen-in-helium mixture, the absolute change is only 1%. Therefore, the increase in the slope at high nitrogen concentrations enables smaller concentration changes to be detected when data at high relative pressures are required.

Figure 14-6 was prepared by flowing helium through one detector while varying the concentration through the second detector.

4. Design Parameters for Continuous Flow Apparatus

The thermal conductivity (T.C.) detector consists of four filaments embedded in a stainless steel or brass block which acts as a heat sink. The T.C. detector is extremely sensitive to temperature changes and should be insulated to prevent temperature excursions during the time that it takes to complete an adsorption or desorption measurement. Long-term thermal drift is not significant, because of the calibration procedure discussed in the next section, and therefore, thermostating is not required.

Figure 14-7 shows a cross-sectional view of a T.C. block and the

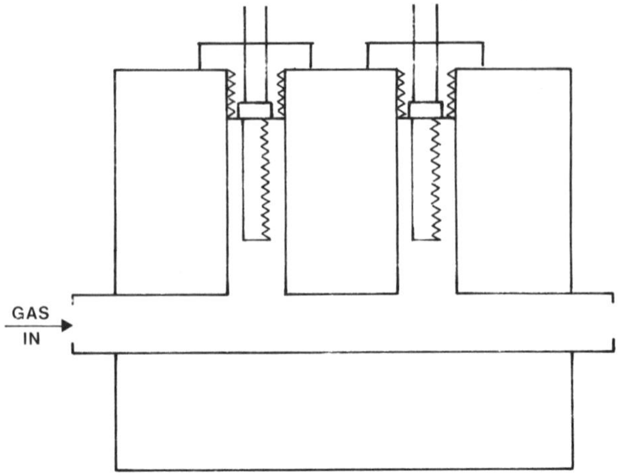

Figure 14-7 Thermal conductivity block with filaments located out of the flow path.

arrangement of the filaments relative to the flow path. The filaments shown are electrically connected, external to the block, and constitute one of the two detectors.

The filaments must be removed from the flow path, unlike the conventional "flow over" type used in gas chromatography, because of the extreme flow variations encountered when the sample cell is cooled and subsequently warmed. Flow variations alter the steady state heat transport from the filaments, leaving them inadequate time to recover before the concentration change from adsorption or desorption is swept into the detector. When this occurs the base line from which the signal is measured will be unstable. By removing the filaments from the flow path and allowing diffusion to produce the signal, the problem of perturbing the filaments is completely solved. However, the trade-off is nonlinear response characteristics. Since the thermal conductivity of the gas mixture is not linear with concentration, this additional nonlinearity poses no additional problems and is accommodated by calibrating the signals, which is the topic of the next section.

A suitable electronic circuit to provide power to the filaments, a means of zeroing or balancing the T.C. bridge, adjusting the filament current, attenuating the signal, and adjusting the polarity is shown in Figure 14-8. Signals produced by adsorption or desorption can be fed to a strip chart recorder for a continuous record of the process and to a digital integrator for summing the area under the adsorption and desorption curves.

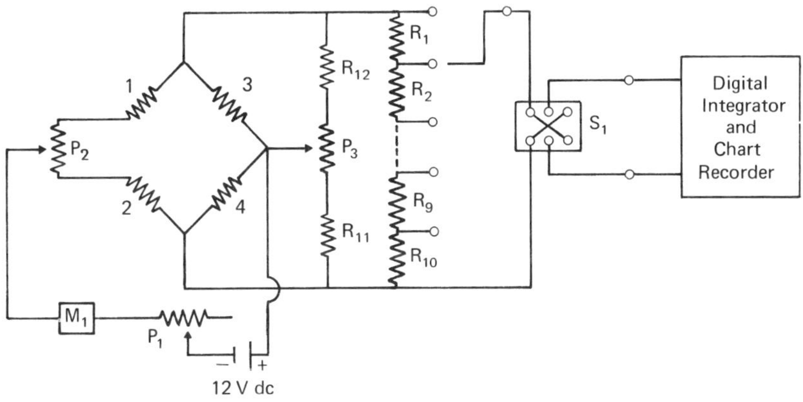

Figure 14-8 Thermal conductivity bridge electronic circuit. 12 V dc power supply, stable to 1 mV. Ripple is not significant due to thermal lag of the filaments. P_1, 100 ohm for filament current control; M_1, milliamp meter, 0–250 mA; P_2, 2 ohms for coarse zero. Filaments 1 and 4 are detector 1; 2 and 3 are detector 2. P_3, 1 ohm for fine zero R_{11}, R_{12}, Padding resistors ~64 ohms; R_1–R_{10}, attenuator resistors 1, 2, 4 . . . 512 ohms; S_1, (D.P.D.T.) switch for polarity. Attenuator resistors are $\frac{1}{4}\%$ ww, lowest temperature coefficient; all others are 1%.

Figure 14-9 is the flow schematic for a commercial continuous flow apparatus. Four gas concentrations from premixed tanks can be selected by valve V_s. Alternatively, gases can be blended internally by controlling their flows with needle valves N_1 and N_2. A third choice is to feed mixtures to the apparatus from two linear mass flow controllers. Flow meters M_1 and M_2, operating under pressure to extend their range, indicate input flow rates. Bubble flow meter B is used to calibrate the flow meters; however, the bubble meter need not be calibrated since it is used to establish flow ratios when blending gases internally. Gauge, G_1 (0–60 psig) indicates the input pressure under which the flow meters operate. Needle valves N_1 and

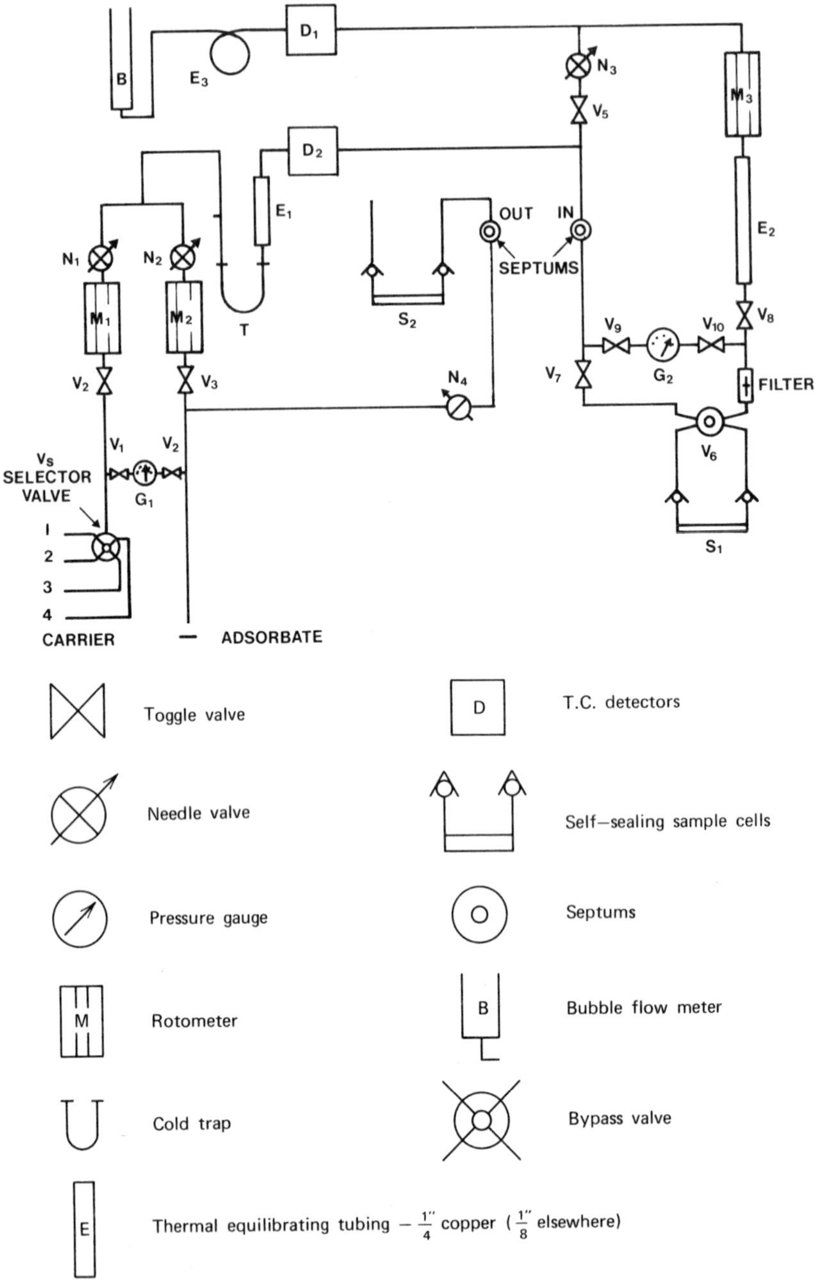

Figure 14-9 Flow diagram for continuous flow system. Courtesy of Quantachrome Corporation.

N_2 control the input flow rates. Cold trap T removes contaminants from the gas stream. After leaving the cold trap but before flowing into detector D_1, the gas passes through wide-diameter tubing which reduce the linear flow velocity and provides time for warming to ambient temperature.

Upon leaving D_1 the flow is split, the larger flow goes to sample cell S_1, through filter F, a thermal equilibrating tube, and a flow meter that indicates the flow through the sample cell. The smaller flow, controlled by needle valve N_3 merges with the sample cell effluent before entering detector D_2. The length of wide-bore tubing E_3 serves as a ballast to prevent air from entering D_2 when the sample is immersed in the coolant.

Splitting the flow serves as a means of diluting the adsorption and desorption concentration peaks in order to continuously adjust the signal height using needle valve N_3. The electronic attenuator shown in Figure 14-8 adjusts the signal by factors of two for each step.

Needle valve N_4 allows a slow bleed of adsorbate to the OUT septum and then to a sample cell positioned at the outgassing station. The slow flow of gas through this part of the circuit provides a source of adsorbate for the purpose of signal calibration and as a purge for the outgassing station holding cell S_2.

The septum labeled IN is used to inject known quantities of adsorbate into the flow to simulate a desorption signal for the purpose of calibration.

In order to measure the saturated vapor pressure, the system is purged with pure adsorbate. Toggle valves V_9 and V_{10} are opened to purge gauge G_2. With toggle valves V_9 and V_8 closed, needle valve N_3 is adjusted until gauge G_2 reads 20–25 torrs. This is sufficient pressure, above ambient, to liquefy nitrogen in the cell when immersed in a liquid nitrogen bath. When V_7 is closed, the sample cell and gauge G_2 are isolated and the gauge reading when added to ambient pressure will be the saturated vapor pressure P_0.

Valve V_6 is rotated to maintain flow continuity when the sample cell is removed.

In order to avoid contaminating the outgassed sample when transferring from the outgassing to the analysis position, the cells are mounted in self-sealing holders that close when disconnected and open when placed in position.

Sample cells consist of a wide variety of designs for various application. Figure 14-10 illustrates seven cells used for various types of samples. Their specific applications and limitations are given in more detail in Section 7 of this chapter. Each of the cells shown are made of Pyrex glass. They are easily filled and cleaned. The cells range 4–5 inches in length with stem inside and outside diameters of 0.15 and 0.24 inches, respectively.

5. Signals and Signal Calibration

The signal intensity created by an adsorption or desorption peak passing through the detector is dependent upon the attenuator setting, the filament current, and the design of the T.C. detector. Also, as stated previously, the detector response is nonlinear. These circumstances require that the adsorption or desorption signals be calibrated by introducing a volume of carrier or adsorbate gas into the flow stream. An expeditious and accurate method of calibration is to withdraw a sample of adsorbate from the OUT septum (see Figure 14-9) with a precision gas syringe and inject a known volume into the flow stream through the IN septum.

Usually the desorption peak is calibrated because it is free of tailing. By immersing the cell in a beaker of water immediately after removal from the liquid nitrogen, the rate of desorption is hastened. Heat transfer from the water is more rapid than from air, therefore, a sharp desorption peak is generated. The calibration signal should be within 20% of the desorption signal height in order to reduce detector nonlinearity to a negligible effect.

The area under the desorption peak A_d and the area under the calibration peak A_c are used to calculate the volume V_d desorbed from the sample according to equation [14.8]

$$V_d = \frac{A_d}{A_c} V_c \qquad [14.8]$$

where V_c is the volume of adsorbate injected. Signals can be plotted

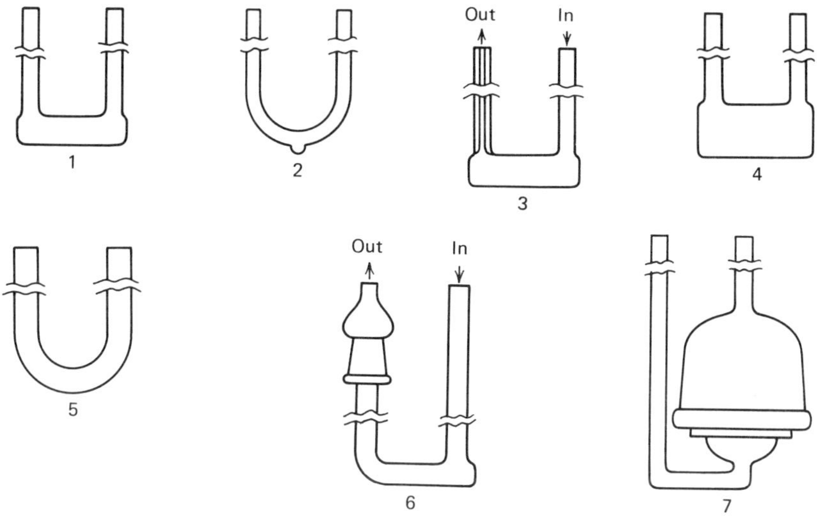

Figure 14-10 Sample cell designs

1. Conventional Sample Cell—for most powder samples with surface areas greater than 0.2 m^2 in the cell. For samples less than 0.2 m^2 this cell can be used with krypton adsorbate.
2. Micro Cell—used for very high surface area samples or for low-area samples that exhibit thermal diffusion signals. Because of the micro cell's small capacity, low-area samples must run on high-sensitivity setting.
3. Capillary Cell—useful for minimizing thermal diffusion when large quantity of low-area sample is required.
4. Macro Cell—used with krypton when a large quantity of sample is required. Also used for chemisorption when specific uptake is small.
5. Large U-Tube Cell—for larger particles or bulk samples of high area with nitrogen or low area with krypton.
6. Pellet Cell—used for pellets or tablets. High surface area with nitrogen or low-area samples with krypton.
7. Monolith Catalyst Cell—Developed for monolithic catalysts and other samples of wide diameter that must be measured as one piece.

on a chart recorder and their areas can be obtained by commercial digital integrators of the type used in gas chromatography (see Figure 14-8). Alternatively, the areas can be measured by planimetry, counting boxes, or by cutting and weighing. The latter methods, however, are less accurate and slower than modern digital integrators.

Equation [14.8] requires no correction for ideality since the volume desorbed is measured at ambient temperature and pressure. Because desorption occurs at room temperature it is complete and represents exactly the quantity adsorbed. For vapors adsorbed near room temperature the sample can be heated to insure complete desorption.

A detailed analysis of the chart record, shown in Figure 14-11, corresponding to a complete adsorption, desorption, calibration, and concentration change cycle discloses that at point *a* the sample cell is immersed in the coolant which produces the adsorption peak P_1. At *b*, the polarity is reversed and the base line is reestablished. Point *c* represents removal of the coolant bath which leads to the desorption peak P_2. The calibration peak P_3 results from the calibration injection made at point *d*. At *e*, a new gas concentration is admitted into the apparatus, which produces a steady base line at *f*, where the detector is rezeroed and the cycle is repeated. The total time for a cycle is usually about 15 minutes. Some time savings can be gained by combining the purge step (*e*–*f*) with the adsorption step (*a*–*b*). Familiarity with the apparatus usually allows the operator to choose the correct volume for calibration at particular attenuator and filament current settings or, alternatively,

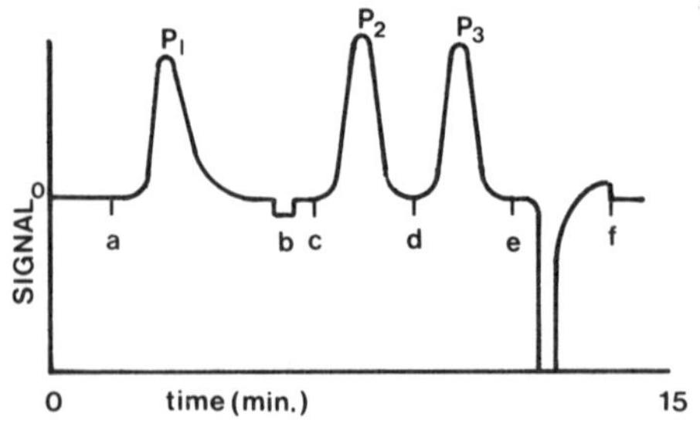

Figure 14-11 Complete cycle for one data point.

a calibration table can be prepared. If speed is essential the flow rate can be increased to hasten the cycle. However, this is at the risk of warming the adsorbent if excessive flow rates are used. Flow rates of 12–15 cm^3 min^{-1} allow ample time for the gas to thermally equilibrate before reaching the sample powder with all the cells shown in Figure 14-10. The presence of helium, with its high thermal conductivity, insures rapid thermal equilibrium. Therefore, immersion depth of the cell is not critical provided that about two and one-half inches is in the coolant at the flow rates given previously.

Sample cells are not completely filled with powder; room is left above the surface for the unimpeded flow of gas. Although the gas flows over the powder bed and not through it, lower flow rates aid to insure against elutriation.

The areas under the adsorption and desorption peaks are usually not exactly the same. This is due to the changing slope of Figure 14-6. Adsorption produces concentration changes to the right or in the direction of decreased sensitivity, while desorption produces signals in the direction of increased sensitivity.

If calibration of the adsorption signal is desired, it is necessary to inject a known quantity of helium. The amount of helium used to calibrate the adsorption signal will usually vary considerably from the amount of nitrogen required for the desorption calibration. This situation arises because, for example, if 1 cm^3 of nitrogen is adsorbed from a 10% flowing mixture, it will produce the equivalent of 9.0 cm^3 of helium. Therefore, calibration of the adsorption signal will require nine times more helium than the corresponding volume of nitrogen needed to calibrate the desorption signal. If C_{N_2} and C_{He} are the concentrations of nitrogen and helium in the flow stream and if V_{He} is the volume of helium used for calibration, then the volume of nitrogen adsorbed V_{ads} is given by

$$V_{ads} = V_{He}\left(\frac{C_{N_2}}{C_{He}}\right)\frac{A_{ads}}{A_{cal}} \qquad [14.9]$$

where A_{ads} and A_{cal} are the areas under the adsorption and calibration signals, respectively.

When small signals are generated, it is difficult to make accurate injections of the required small amounts of gas. Karp and Lowell[109] have shown that this problem can be solved by injecting larger volumes of adsorbate diluted with the carrier gas. When a volume containing a mixture of nitrogen and helium V_{mix} is injected into the flow stream, the equivalent volume of pure nitrogen V_{N_2} is given by

$$V_{N_2} = V_{mix}\left(\frac{X_{N_2} - X'_{N_2}}{X'_{He}}\right) \qquad [14.10]$$

where X_{N_2} is the mole fraction of nitrogen in the calibration—mixture, X'_{N_2} and X'_{He} are the mole fractions of nitrogen and helium in the flow stream.

The signals must propagate through the system at identical flow rates. Calibration at a flow rate other than the flow rate associated with the adsorption or desorption peaks can lead to serious errors because the width of the peaks, and therefore the peak areas, are directly proportional to the flow rate. A good two-stage pressure regulator and needle valve will provide adequately constant flow rates over the short time required for desorption and calibration.

Precision gas calibrating syringes can be obtained in various size ranges with no more than 1% volumetric error. Constant stroke adapters provide a high degree of reproducibility. Often, in the BET range of relative pressures, the calibration volumes remain nearly constant because the increased volume adsorbed at higher relative pressures tends to be offset by the decrease in the detector sensitivity. Thus, the same syringe may be used for a wide range of calibrations, which results in the syringe error not effecting the BET slope and only slightly altering the intercept which usually makes a small contribution to the surface area. A syringe error of 1% will produce an error in surface area far less than 1% for those BET plots with a slope greater than the intercept or for high C values.

6. Adsorption and Desorption Isotherms by Continuous Flow

To construct the adsorption isotherm, the adsorption, desorption and calibration cycle shown in Figure 14-11 is repeated for each data point required. Errors are not cumulative since each point is independently determined. Relative pressures corresponding to each data point are established by measuring the saturated vapor pressure using any of the aforesaid methods or by adding 15 torrs to the ambient pressure; thus if X is the mole fraction of adsorbate in the flow stream the relative pressure is given by

$$\frac{P}{P_0} = \frac{XP_a}{P_a + 15} \qquad [14.11]$$

where P_a is ambient pressure in torrs. At the recommended flow rates of 12–15 $cm^3\ min^{-1}$ the flow impedance of the tubing does not raise the pressure in the sample cell.

The method used to construct the adsorption isotherm cannot be used to build the desorption isotherm. This is because each data point on the adsorption curve reflects the amount adsorbed by a surface initially free of adsorbate. The desorption isotherm, however, must consist of data points indicating the amount desorbed from a surface that was previously saturated with adsorbate and subsequently equilibrated with adsorbate of the desired relative pressure. Karp, Lowell, and Mustacciuolo[110] have demonstrated that the desorption isotherm and hysteresis loop scans can be made in the following manner. First, sample is exposed while immersed in the coolant to a flow of pure adsorbate. The flow is then changed to the desired concentration, leading to some desorption until the surface again equilibrates with the new concentration. The coolant is next removed and the resulting desorption signal is calibrated to give the volume adsorbed on the desorption isotherm. The preceding procedure is repeated for each data point required, always starting with a surface first saturated with pure adsorbate.

To scan the hysteresis loop from the adsorption to the desorption isotherm, the sample, immersed in the coolant, is equilibrated with a gas mixture with a relative pressure corresponding to the start of the scan on the adsorption isotherm. The adsorbate concentration is then reduced to a value corresponding to a relative pressure between the adsorption and desorption isotherms. When equilibrium is reached, as indicated by a constant detector signal, the coolant is removed and the resulting desorption signal is calibrated. Repetition of this procedure, each time using a slightly different relative pressure between the adsorption and desorption isotherms, yields a hysteresis scan from the adsorption to the desorption isotherm.

To scan from the desorption to the adsorption branch, pure adsorbate is first adsorbed, then the adsorbate concentration is reduced to a value giving a relative pressure corresponding to the start of the scan on the desorption isotherm. When equilibrium is established as indicated by a constant base line, the adsorbate concentration is increased to give a relative pressure between the desorption and adsorption isotherms. After equilibrium is again established, the coolant is removed and the resulting signal is calibrated to yield a data point between the desorption and adsorption isotherms. This procedure repeated, each time using a different final relative pressure, will yield a hysteresis loop scan from the desorption to the adsorption isotherm.

Figures 14-12 and 14-13 illustrate the results obtained using the preceding method on a porous amorphous alumina sample. A distinct advantage of the flow system for these measurements is that data points can be obtained where they are desired and not where they happen to occur after dosing, as in the vacuum volumetric

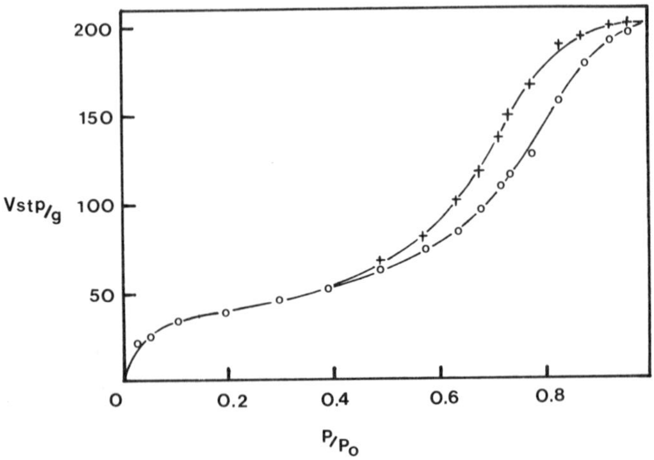

Figure 14-12 Adsorption and desorption isotherms of N_2 for 0.106g sample of alumina. Adsorption, ○; desorption, +. V in cm^3.

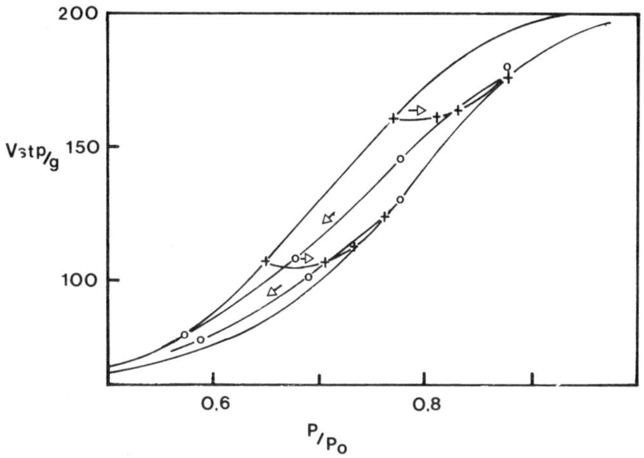

Figure 14-13 Hysteresis loop scan for same sample as Figure 14-12. Desorption to adsorption, + ; adsorption to desorption, ○.

method. In addition, desorption isotherms and hysteresis scans are accomplished with no error accumulation, void volume, or ideality corrections.

7. Low Surface Area Measurements

The thermal conductivity bridge and flow circuits shown in Figures 14-8 and 14-9 are capable of producing a full-scale signal (1.0 mV) when 0.001 cm^3 of nitrogen is desorbed into a 30% nitrogen and helium mixture. To achieve stable operating conditions at this sensitivity the thermal conductivity block will require about one hour to thermally equilibrate and the system must be purged of any contaminants.

A desorbed volume of 0.001 cm^3, using nitrogen as the adsorbate, will correspond to about 0.0028 m^2 (28 cm^2) of surface area if a single adsorbed layer were formed. An equivalent statement is that 0.0028 m^2 is the surface area, measured by the single point method, on a sample that gives a high C value, if 0.001 cm^3 were desorbed. Assuming that a signal 20% of full scale is sufficient to give reasonable accuracy for integration, then the lower limit for surface area measurement using hot wire detection is about 0.0006 m^2 or 6 cm^2. With thermistor detectors the lower limit would be still smaller.

Long before these extremely small areas can be measured with nitrogen, the phenomenon of thermal diffusion obscures the signals and imposes a higher lower limit.[111] Thermal diffusion produces a tendency for a gas mixture to separate when exposed to a temperature gradient. The arms of a sample cell experience a steep temperature gradient where they enter the liquid nitrogen bath. The resultant effect is for the heavier molecule to migrate to the high-temperature region. In a static mixture of gases the amount of thermal diffusion is a function of the temperature gradient, the gas concentration, and the difference in masses of the molecules. In a flowing gas mixture, in the presence of adsorption, it is difficult to assess the exact amount of thermal diffusion.

Lowell and Karp[112] measured the effect of thermal diffusion on surface areas using the continuous flow method. Figure 14-14 illustrates a fully developed anomalous desorption signal caused by thermal diffusion.

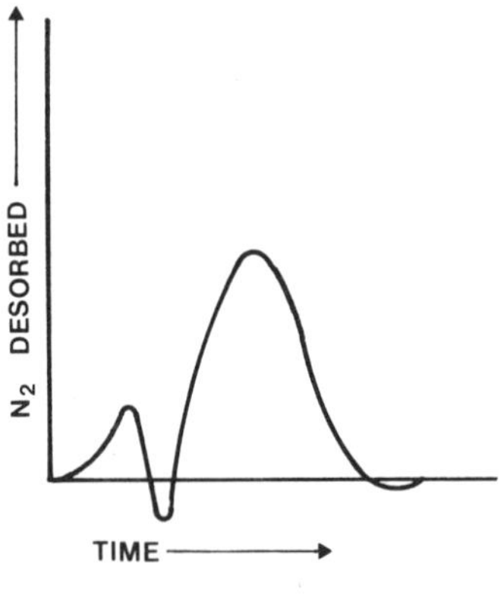

Figure 14-14 Desorption of a small volume of nitrogen.

As a result of the positive and negative nature of the signal, accurate integration of the true desorption peak is not possible. Table 14-1 shows the results of measuring the surface area of

Table 14-1. Data Obtained Using Conventional Cell[a]

Weight	Actual Area (m^2)	Measured Area (m^2)	Deviation (%)	Signal Shape at Start of Desorption Peak
1.305	5.07	5.07	0	
0.739	2.87	2.87	0	
0.378	1.47	1.45	−1.3	
0.177	0.678	0.686	−0.14	
0.089	0.345	0.327	−5.5	
0.049	0.190	0.166	12.6	
0.0190	0.0730	0.0481	34.1	
0.0101	0.0394	0.0192	51.3	

[a]Measured using the one-point BET method with 20% N_2 in He.

various quantities of zinc oxide using a conventional sample cell, Number 1, in Figure 14-10.

When the same sample was analyzed using a micro cell, number 2 in Figure 14-10, the results obtained were considerably improved, as shown in Table 14-2.

The onset of thermal diffusion depends upon the gas concentrations, the sample surface area, the rate at which the sample cools to

Table 14-2. Data Obtained Using U-Tube Cell

Wt ZnO (g)	Actual Area (m^2)	Measured Area (m^2)[a]
0.059	0.228	0.230
0.027	0.105	0.102
0.0045	0.0175	0.0161

[a]This value is less 15 cm^2 for the cell wall area, as estimated from the cell dimensions. Desorption peaks from an empty U-tube cell gave areas of 12–17 cm^2.

bath temperature, and the packing efficiency of the powder. In many instances, using a conventional sample cell, surface areas less than 0.1 m can be accurately measured on well-packed samples that exhibit small interparticle void volume. The use of the micro cell (Figure 14-10, 2) is predicated on the latter of these observations. Presumably, by decreasing the available volume into which the lighter gas can settle, the effects of thermal diffusion can be minimized. Although small sample quantities are used with a micro cell, thermal conductivity detectors are sufficiently sensitive to give ample signal.

Another cell design that aids to minimize the effects of thermal diffusion is the capillary cell, Number 3, in Figure 14-10. By using capillary tubing on the vent side of the cell, a sufficiently high flow rate is maintained to prevent that arm from contributing to the problem. The large sample capacity of the capillary cell, compared to the micro cell, produces sufficient desorption signal to often make the thermal diffusion effect negligibly small.

Lowell[113] has published a method to circumvent the problem of thermal diffusion by using an adsorbate with a low vapor pressure such as krypton at liquid nitrogen temperature. The coefficient of thermal diffusion $D(t)$ is given by[114]

$$D(t) = \frac{N_1/N_t - N'_1/N_t}{T_1/T_2} \qquad [14.12]$$

where N_1 and N'_1 are the adsorbate concentrations at the absolute temperatures T_2 and T_1, respectively, with $T_2 > T_1$. The term N_t is the total molecular concentration of adsorbate and carrier gas.

Because of krypton's low vapor pressure its mole fraction in the BET range of relative pressures is of the order of 10^{-4} atm. This small value causes the difference between N_1/N_t and N'_1/N_t to nearly vanish, with the consequence that no obscuring thermal diffusion signals are generated.

Attempts to increase the size of nitrogen adsorption or desorption signals, by using larger sample cells, result in enhanced thermal diffusion signals due to the increased interparticle void volume into which the helium can settle. However, when krypton is used, no thermal diffusion effect is detectable in any of the sample cells shown in Figure 14-10.

The adsorption signals using krypton–helium mixtures are broad and shallow because the adsorption rate is limited by the low vapor pressure of krypton. The desorption signals are sharp and comparable to nitrogen since the rate of desorption is governed by the rate of heat transfer into the powder bed.

With krypton, the ability to use larger samples of low-area powders facilitates measuring low surface areas because larger signals are generated in the absence of thermal diffusion. Also, as is true for nitrogen, krypton measurements do not require void volume, or ideality corrections, nor is thermal transpiration a factor as in the volumetric measurements.

8. Data Reduction—Continuous Flow

Table 14-3 can be used as a work sheet for calculating specific surface area from continuous flow data. The first five rows in the lower left corner are entered first and are used to calculate other entries. In the example shown, nitrogen is the adsorbate.

Column 1 is the mole fraction of adsorbate in the flow stream. Column 2 is obtained as the product of P_a and column 1. Column 2 when divided by column 3 gives the relative pressure, which is entered in column 4 and from which columns 5 and 6 are calculated.

Column 7 is the volume required to calibrate the desorption signal and column 8 is the corresponding weight of the calibration injection. Column 8 is calculated from the equation in the lower

Figure 14-3. Multipoint BET Surface Area Data Sheet

Date ______________ Sample ______________ Total Weight 11.5614

Operator ______________ Outgas Procedure He purge, 1 hr, 150°C Tare 11.3918

Sample Weight (M) 0.1696

1	2	3	4	5	6	7	8	9	10	11	12	13
X_{N_2}	P (torr)	P_0 (torr)	P/P_0	P_0/P	$(P_0/P)-1$	V_c (cm^3)	W_c (g)	A_s	A_c	$W=\frac{A_s}{A_c}W_c$	$W(P_0/P-1)$	$\frac{1}{W(P_0/P-1)}$
0.107	80.69	769	.1049	9.532	8.532	2.00	.00231	1108	1080	0.00267	0.0228	43.84
0.185	139.5	769	.1814	5.513	4.513	2.60	.00300	1186	1155	0.00308	0.0139	71.94
0.246	185.5	769	.2412	4.146	3.146	3.00	.00347	1182	1218	0.00337	0.0106	94.34
0.317	239.0	769	.3107	3.219	2.219	3.30	.00381	1008	1027	0.00374	0.0083	120.5

Ambient Pressure, P_a 754.1 mm Hg

Vapor Pressure, P_0 769.0 mm Hg

Ambient Temperature, T 293°K

Adsorbate Molecular Weight, $\overline{M}$ 28.01

Adsorbate Area, $\mathscr{A} = 16.2 \times 10^{-20}$ m^2

Calibration Gas Weight $W_c = \dfrac{P_a M V_c}{(6.235 \times 10^4)T}$

Plot $\dfrac{1}{W[(P_0/P)-1]}$ vs P/P_0

Slope, $s = 367.6$

Intercept, $i = 4.57$

$$W_m = \frac{1}{s+i}$$

$W_m = 0.00269$ g

Total Surface Area,

$$S_t = \frac{W_m(6.023 \times 10^{23})}{\overline{M}} \mathscr{A}$$

$S_t = 9.36$ m^2

$S = S_t/W$

$S = 55.19$ m^2 g^{-1}

left side of the work sheet. The terms A_s and A_c are the areas under the signal and calibration peaks, respectively.

Columns 11–13 are calculated from the data in the previous columns. The data in column 13 is then plotted versus the corresponding relative pressures in column 5. The slope s and intercept i are calculated, and the value of W_m is found as the reciprocal of their sum. Equation [5.13] is used to obtain the sample surface area S_t and dividing by the sample weight yields the specific surface area.

9. Single Point Method

The assumption of a zero intercept reduces the BET equation to equation [6.3]. This assumption is, of course, not realizable since it would require a BET C value of infinity. Nevertheless, many samples possess sufficiently high C values to make the error associated with the single point method acceptably small, (see Chapter 6 and Table 6-1).

Using the zero intercept assumption, the BET equation can be written as

$$W_m = W\left(1 - \frac{P}{P_0}\right) \qquad \text{[cf. 6.3]}$$

from which the total surface area can be calculated by

$$S_t = W\left(1 - \frac{P}{P_0}\right)\frac{\overline{N}}{\overline{M}}\mathscr{A} \qquad \text{[cf. 6.4]}$$

From the ideal gas equation of state

$$W = \frac{P_a V \overline{M}}{RT} \qquad [14.13]$$

so that

$$S_t = \left(1 - \frac{P}{P_0}\right)\frac{P_a V \overline{N}}{RT}\mathscr{A} \qquad [14.14]$$

where as previously P_a and T are the ambient pressure and absolute temperature, respectively, $\bar{N}$ is Avogadro's number, $\mathscr{A}$ is the adsorbate cross-sectional area, and V is the volume adsorbed.

Using nitrogen as the adsorbate and the recommended relative pressure of 0.3, (see Chapter 6, Section 3), equation [14.14] reduces to

$$S_t = (2.84)V \text{ m}^2 \qquad [14.15]$$

Thus, the total surface area contained in the sample cell is given by the simple linear relationship above when V is in cubic centimeters.

By calibrating the desorption signal A_d with a known volume of nitrogen V_c, equation [14.15] can be rewritten as

$$S_t = (2.84)\frac{A_d}{A_c}V_c \qquad [14.16]$$

where A_d and A_c are the integrated areas under the desorption and calibration signals, respectively.

A commercial single point instrument[115] contains a linearization network which corrects for the hot wire detector nonlinearity. This procedure allows a built-in digital integrator to linearly integrate the signals so that the surface area is given directly on a digital display. Analysis time on this apparatus is extremely short, usually under 10 minutes.

CHAPTER FIFTEEN

Other Flow Methods

1. Pressure Jump Method

A modification of the Nelson and Eggertsen method was made by Haley,[116] who also used helium and nitrogen in a continuous flow system. Haley's apparatus differs from that described in the previous chapter in that the sample cell can be pressurized while the effluent flows through a pressure regulator to the detector which is maintained at ambient pressure.

Starting with a 10% nitrogen-in-helium mixture, the sample, held slightly above ambient pressure, is immersed in the bath. When the adsorption signal is complete, the pressure is raised to 2 atm, which increases the partial pressure of nitrogen to 0.2 atm or the equivalent of a 20% mixture at 1.0 atm. After the completion of each adsorption signal, the pressure is incrementally increased until, at 150 psia, saturation is reached.

Desorption is accomplished by incrementally decreasing the pressure and monitoring the nitrogen-rich desorption signals.

Haley used a gas calibration loop of fixed volume to calibrate the signals. Under pressures as high as 150 psia septums and syringes are prone to leakage and injections of large volumes are difficult to achieve due to the opposing pressure. Also, the detector's nonlinearity requires that the calibration signal closely match the adsorption or desorption signal so that a fixed volume calibration loop is questionable in this application.

The entire apparatus must be very tight to prevent leakage at high pressures. The sample cell should be metal or heavy wall glass and is wrenched into position. A further difficulty is that at high pressures the linear flow velocity through the sample cell is small. For example, to maintain the same volumetric flow rate through the detector at 150 psia requires one-tenth the linear flow rate at 15 psia because of gas expansion to atmospheric pressure. Therefore, the signal duration is excessively long, being broader and shorter as the pressure increases. Increasing the flow rate makes the signals narrower but their height remains small due to the dilution within the sample cell at higher pressures.

2. Continuous Isotherms

Semonian and Manes[117] have devised an approach that provides continuous data from which the desorption isotherm can be constructed. Their method utilizes a calibrated thermal conductivity detector for sensing the effluent concentration from a cell filled with adsorbate and slowly purged with a carrier gas. The amount desorbed at any relative pressure is calculated by integrating the effluent flow rate and thermal conductivity signal.

The tail end of the signal approaches the zero signal axis asymptotically, making accurate integration difficult and extending the time required for analysis. This difficulty is overcome by allowing the cell temperature to rise in order to hasten the desorption rate at the tail end of the curve. Using the Polanyi potential theory, Semonian and Manes reduced the nonisothermal data to points on the isotherm by equating the adsorption potential at various temperatures to the potential at the isotherm temperature, viz.,

$$RT \ln \frac{P_0}{P} = RT' \ln \frac{P'_0}{P'} \qquad \text{[cf. 11.2]}$$

The partial pressure P' at temperature T' is calculated from P and T, the measured values.

The method gave good agreement with static data using *n*-butane as the adsorbate, nitrogen as the carrier gas, and activated carbon as the adsorbent.

3. Frontal Analysis

Early theories of elution chromatography developed by Wilson,[118] Glueckauf,[119-121] De Vault,[122] and others,[123-124] established the theoretical foundations for the development and propagation of solute bands through a packed column. Stock[125] succeeded in testing these theories for the elution of volatile adsorbed substances from a packed column by an inert gas. Stock measured the effluent concentration using a thermal conductivity detector and constructed the isotherms from the resulting "chromatograms." Gregg[126] outlined the method giving several examples and an explanation of the theory. The method of frontal analysis has been used and discussed by Cremer and Huber.[127] Malamud, Geisman and Lowell[128] have measured the isotherms of *o*-, *m*-, and *p*-xylene isomers on zinc oxide using frontal analysis.

The following mathematical description[126] of the development of a chromatographic elution profile assumes that the vapor is in equilibrium with the adsorbent, that diffusion is absent or of negligible rate compared to the flow rate, and that the adsorbent is uniformly packed in the column.

Consider that the flow through a packed column is suddenly switched from pure carrier gas to a mixture of carrier gas and adsorbate vapor at concentration c. As the mixture traverses the distance dl along the column, adsorption will cause the concentration to decrease by dc. Expressing the concentration as weight W of adsorbate per unit volume of mixture, gives the weight lost by the flow in the column length dl as

$$dW = \left(\frac{\partial c}{\partial l}\right)_v dl\,dv \qquad [15.1]$$

where dv is the volume of the mixture which has traversed the distance dl.

The weight gained by the adsorbent in the length dl is

$$dW = \left(\frac{\partial W}{\partial v}\right)_l dv dl \qquad [15.2]$$

Then, from equations [15.1] and [15.2]

$$\left(\frac{\partial W}{\partial v}\right)_l = \left(\frac{\partial c}{\partial l}\right)_v \qquad [15.3]$$

In general, for any isotherm

$$W = f(c) \qquad [15.4]$$

and

$$dW = f'(c)\, dc \qquad [15.5]$$

then

$$f'(c) = \left(\frac{\partial c}{\partial l}\right)_v \left(\frac{\partial v}{\partial c}\right)_l = \left(\frac{\partial v}{\partial l}\right)_c \qquad [15.6]$$

Taking the reciprocal of the preceding yields

$$\left(\frac{1}{f'(c)}\right) = \left(\frac{\partial l}{\partial v}\right)_c \qquad [15.7]$$

Equation [15.7] asserts that a volume of the mixture of adsorbate and carrier gas, at concentration c, moves through the column at a rate that is inversely proportional to the slope of the isotherm at that concentration.

Isotherms of type I are characterized by a continuously decreasing slope which often becomes zero. Therefore, the adsorbate will move more rapidly through the column at higher concentrations. When the flow is switched from a mixture to pure carrier gas, the elution profile or chromatogram will appear as shown in Figure 15-1 when plotted on a strip chart recorder.

In Figure 15-1, P is the signal produced by saturated vapor, t is the time of switch-over to pure carrier gas. At time t_2 the signal corresponds to the pressure P and the horizontally shaded portion

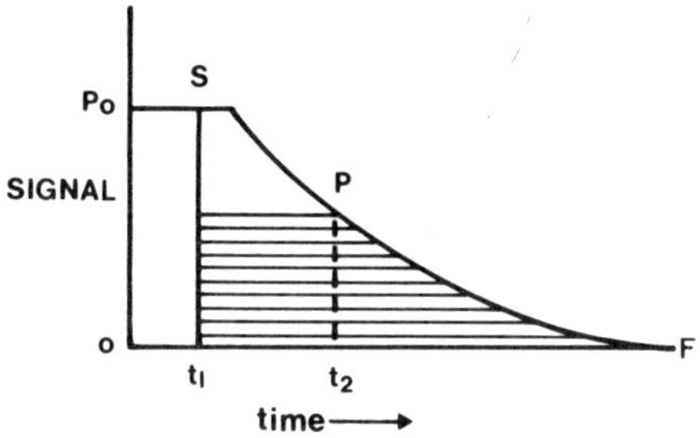

Figure 15-1 Elution profile for type I isotherms.

of the curve is proportional to the quantity of adsorbate remaining in the column. By successive integrations of the area under the curve along the line *SPF*, the quantity adsorbed at various relative pressures can be calculated.

Normalization of the signal is achieved by measuring the signal height produced by saturated vapor of known vapor pressure in the carrier gas. The area produced on the chart record for a constant flow rate over a known time interval at saturation generates all the information required for calibration. To reduce a chromatogram to the corresponding isotherm, the weight adsorbed per unit weight of adsorbent, at a particular vapor pressure, is related to the chromatogram signal area by equation [15.8]

$$\frac{W}{M}=\frac{PF\bar{M}A}{SCTRP_a} \qquad [15.8]$$

where

W = grams adsorbed
M = grams of adsorbent in the column
$\bar{M}$ = molecular weight of adsorbate
A = area under the signal curve at a given pressure (sq. inches)
F = flow rate ($cm^3\ sec^{-1}$)
P = adsorbate vapor pressure (mm Hg)
S = signal height when $P/P_0 = 1$ (inches)
C = chart speed (inches min^{-1})
T = absolute temperature (°K)
R = gas constant, 82.1 ml atm deg^{-1} $mole^{-1}$
P_a = standard pressure, 760 mm Hg

The values of W/M calculated from equation [15.8] must be corrected for the weight of adsorbate contained in the column void volume, which can be obtained by subtracting the powder volume from the volume of the empty column.

Type III isotherms possess a continuously increasing slope. Therefore, the adsorbate propagates more rapidly at lower concentrations. The corresponding elution profile when the flow is switched from pure carrier gas to a mixture of carrier and adsorbate is shown in Figure 15-2.

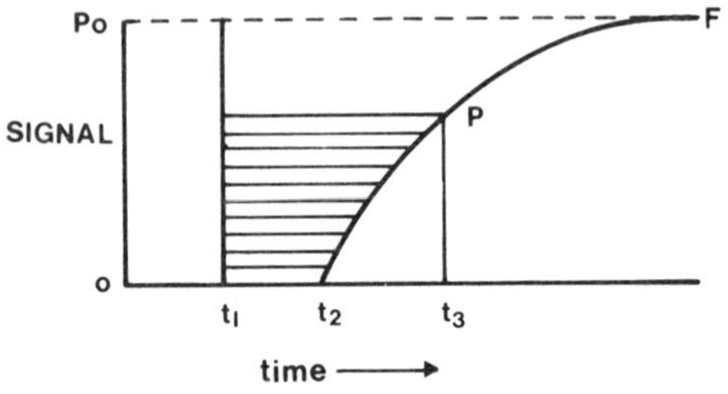

Figure 15-2 Elution profile for type III isotherms.

The times t_1, t_2, and t_3 in Figure 15-2 correspond, respectively, to the time when the flow is switched from carrier to mixture, when the signal begins to emerge, and when the concentration of the adsorbate has risen to pressure P. The horizontally shaded area is proportional to the quantity remaining in the column at pressure P. By successively integrating the areas between the line t_2PF and the saturated pressure signal, the adsorbed quantity can be calculated at various relative pressures. Gregg[126] discusses the development of the types II, IV, and V isotherms which are more complex than the examples given here. However, it is interesting to note that a type II isotherm, for example, can be considered equivalent to a type I at the lower end and a type III at the upper end. Therefore, it is possible to obtain the entire isotherm in sections. By examining the rear of the elution profile after loading the column to about 0.4 relative pressure, the lower part can be obtained while the upper section is determined after charging the column with saturated vapor and examining the front of the elution profile.

For isotherms that exhibit hysteresis this method is of questionable value, since the high relative pressure end is measured as an adsorption isotherm but the low-pressure end is measured as a desorption isotherm.

CHAPTER SIXTEEN

Gravimetric Method

1. Electronic Microbalances

Balances with adequate sensitivity to measure the small weight changes associated with adsorption can be divided into two categories. First are the electronic beam microbalances. These balances usually use an electric current to restore the beam to the horizontal position as the weight changes. Their sensitivities are in the order of 1 μg with loadings of approximately one gram. Commercial electronic microbalances are available in vacuum containers with appropriate fittings for vacuum pumps and gauges.

Figure 16-1 is a rough drawing of the Cahn[129] vacuum microbalance. The sensitivity is 1 μg with a load of 0.5 gram. Counter weight A is used to balance the adsorbent contained in bucket G, which is immersed in a liquid nitrogen filled Dewar flask H. Photo cell C senses movement of the beam and provides current to the coil B to restore the beam. A metal plate D is fastened against the ground glass end of the vacuum bottle with electrical connections for bringing the signal to a recorder in order to continuously monitor the weight changes. Hang-down wire F is made of fine stainless steel. Ground joint E leads to pumps and gauges. For larger weights the inner hook K can be used.

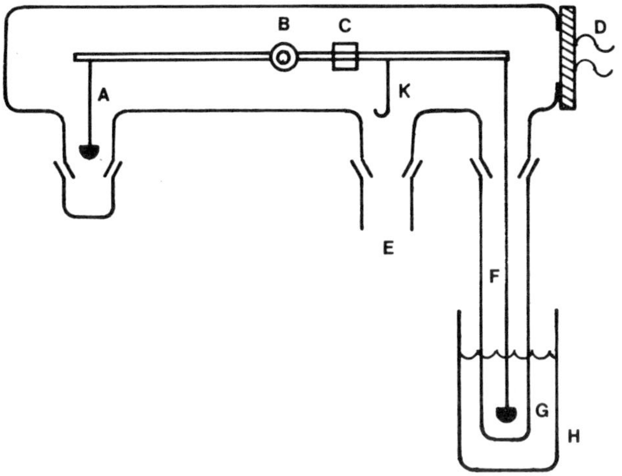

Figure 16-1 Cahn vacuum microbalance.

2. Buoyancy Corrections

When using beam microbalances the effects due to buoyancy and thermal transpiration must be considered. The gas displaced by the sample produces a buoyancy force that reduces the weight by exactly the mass of displaced gas. Buoyancy effects both the mass being weighed and the counterweight. Also, the buoyancy changes with pressure. The mass of gas displaced is given by

$$W = \frac{\bar{M}PV}{RT} \qquad [16.1]$$

Allowing V_w and T_w to be the volume and absolute temperature of the counterweight, V_c and T_c to be the same for the sample and its container, the buoyancy correction becomes

$$\Delta W = \frac{P\bar{M}}{R}\left(\frac{V_c}{T_c} - \frac{V_w}{T_w}\right) \qquad [16.2]$$

To estimate the error introduced by buoyancy using nitrogen as the adsorbate, allow T_w and T_c to equal 295 and 77° K, respectively, and

$V_w = V_c = V$; equation [16.2] then becomes

$$\Delta W = 0.00327PV \qquad [16.3]$$

with P in atmospheres and V in cubic centimeters. In a completed monolayer 1 μg of nitrogen will occupy 34.8 cm^2 of surface. Therefore, if the volumes V_c and V_w are each 2 cm^3, an error of 0.1 torr in pressure measurement will produce a weighing error of 0.86 μg or a surface area error of 29.9 cm. For 1% accuracy it would then be necessary for the sample to exhibit an area not less than 2990 cm^2 (0.299 m^2).

The error caused by pressure measurement is not nearly as significant as that produced by measurements of the sample and counterweight volumes. By evacuating the container and admitting helium to a known pressure, say 300 torrs, the term in parenthesis of equation [16.2] can be calculated. With this information, ΔW for nitrogen can be calculated at any desired pressure. If the buoyancy effect with helium can be evaluated to within 1% then the error, ΔW for nitrogen, assuming V in equation [16.3] is 2.5 cm^3 and P is 200 torrs near the region of monolayer coverage, would be

$$\Delta W = 0.00327\left(\frac{200}{760}\right)(2.5)(0.01) = 21.5\ \mu g \qquad [16.4]$$

The corresponding surface area error would be 748 cm^2. Thus, to insure accuracy of 1% or better it would be necessary to measure no less than about 7.5 m^2.

3. Thermal Transpiration

As stated previously, in Chapter 13, Section 7, thermal transpiration or thermomolecular flow arises when gas is contained in two vessels at equal pressures but different temperatures and a connection is made between the vessels. A vacuum microbalance represents just such a system. The long tube surrounding the hand-down wire separates the sample, immersed in the coolant bath, from the warm upper portion of the apparatus. With vacuum microbalances the effects of thermal transpiration is further exacerbated by the temperature gradient along the hang-down wire.

When using low vapor pressure adsorbates, such as krypton at liquid nitrogen temperature, the entire isotherm lies within the thermal transpiration region. The effect of thermal transpiration is to produce an equilibrium uptake at a pressure less than that indicated on the manometer. The microbalance will indicate a weight change as thermomolecular flow occurs in the range 10^{-3}–20 torrs. The extent of the weight change depends on the physical dimensions of the tube enclosing the sample, the size of the sample, thermal conduction of the hang-down wire, the gas used, and the pressure.

With nitrogen the data required for BET analysis or pore volumes lies above the thermal transpiration region.

When using microbalances for adsorption measurements, those adsorbates that do not require thermal transpiration corrections are the most susceptible to buoyancy errors whereas those adsorbates not requiring buoyancy corrections, such as krypton because of its low vapor pressure, are most susceptible to thermal transpiration errors.

4. Other Gravimetric Methods

Helical spring balances first used by McBain and Bakr[130] have been extensively used for adsorption measurements. The spring is suspended inside a glass tube by attachment to a hook at the top. A bucket containing the adsorbent is connected to the bottom of the spring. The bottom of the tube, containing the sample, is immersed in the coolant. The upper portion of the tube is connected to a vacuum pump, source of adsorbate, and manometer.

The spring is calibrated by adding known weights and measuring the extension with a cathetometer. The upper and lower portions of the tube should be thermostated to retain the spring calibration. Temperature coefficients of extension can amount to 10% of the spring's sensitivity, which is about 1 mg mm^{-1} with silica springs, which are commonly employed due to their linearity, high sensitivity, and reversibility.

Many improvements in design and material have been made since McBain and Bakr's original paper. Gregg and Sing,[14] as well as Young and Crowell,[131] review some of the more recent developments of spring as well as beam balances.

Other gravimetric techniques include flowing a mixture of carrier gas and adsorbent through a tube packed with adsorbent. Stopcocks at each end of the tube permit isolation of the sample. The tube is then removed from the flow stream and weighed. Two successive equal weights indicate that equilibrium is achieved. The quantity adsorbed must be corrected for the gas contained in the void volume.

CHAPTER SEVENTEEN

Comparison of Experimental Adsorption Methods

The following comparisons of the three most common adsorption techniques (volumetric, continuous flow, and gravimetric) are based on the assumption that routine measurements are to be made. Special requirements may oblige the experimenter to choose one method in preference to another.

Ideal Gas Corrections— Volumetric measurements require that adsorbed volumes be corrected for ideality. The continuous flow method does not require this correction because the detector senses the gas concentration change at ambient temperature and pressure. Gravimetric apparatus measures the adsorbed weight directly and needs no correction for deviations from ideality.

Void Volume Measurements— The void volume must be measured in volumetric systems. The continuous flow and gravimetric methods are independent of the void volume.

Buoyancy Corrections—Buoyancy corrections are required only by the gravimetric method.

Speed—Volumetric systems approach equilibrium asymptotically. The heat of adsorption is dissipated primarily through the powder bed at low pressures in the BET range. Continuous flow apparatus are more rapid because a fresh supply of adsorbate is constantly brought to the adsorbent, thereby driving the system to equilibrium. Heat is dissipated rapidly through the powder bed as well as through the flowing gas mixture which is at ambient pressure and contains helium. Gravimetric instruments approach equilibrium asymptotically as does the volumetric apparatus, but heat dissipation tends to be slow, particularly at low pressures where the principal source of thermal conduction is the hang-down wire.

Data Points—The gravimetric and volumetric methods involve dosing the sample with adsorbate; the system subsequently comes to an equilibrium pressure which depends on the dosing volume, the isotherm shape, and the quantity of adsorbent. The continuous flow method produces data at the concentration of adsorbate in the flow stream. Therefore, the exact position of the data point can be chosen.

Thermal Transpiration— Both the volumetric and gravimetric apparatus are subject to this effect. Data acquired at low pressures must be corrected for thermal transpiration. The continuous flow method is not subject to this phenomenon.

Thermal Diffusion—The continuous flow method uses gas mixtures and is, therefore, the only one of the three methods subject to this effect. It occurs only when low areas are measured and can be eliminated by using an adsorbate with low vapor pressure such as krypton.

Liquid Nitrogen Level— Gravimetric and continuous flow equipment require only that the coolant level be sufficiently high to maintain the sample at bath temperature. The volumetric method requires that the level be maintained constant in order to avoid changing the void volume.

Thermostating— For maximum accuracy the manifold and calibrated volumes in a volumetric apparatus should be maintained at constant temperature. Thermostating is not necessary for vacuum microbalances but in helical spring balances the spring should be maintained at constant temperature. Continuous flow apparatus need not be thermostated since the signals are immediately calibrated with known volumes at the same temperature and pressure. However, ambient temperature and pressure must be known to insure accurate calibration.

Gas Mixtures—Only the continuous flow method requires mixed gases. Gas mixtures can be purchased commercially and can be prepared to accuracies of 1% relative. Alternatively, flows can be blended but with often some loss in accuracy.

Vacuum—The gravimetric and volumetric apparatus both require vacuum systems. The continuous flow method does not.

Outgassing—Because the gravimetric and volumetric apparatus use vacuum systems it is convenient to outgas by vacuum. Outgassing in continuous flow apparatus is accomplished by purging.

Calibration—Volumetric systems are calibrated by weighing the amount of mercury required to fill internal volumes. Gravimetric systems are calibrated using reliably known weights and should be recalibrated occasionally to insure that the sensitivity has not drifted with use. Continuous flow instruments are calibrated with gas sampling loops or syringes, the accuracy of which can be determined by filling with mercury and weighing.

Sample Size—Gravimetric systems are limited to the maximum loading of the balance, usually about one gram. Volumetric apparatus should use samples sufficiently large to reduce the void volume error to an acceptable percentage. Usually a sample with about 10 m^2 of surface area is used with nitrogen. Flow systems can accommodate a wide range of sample size and are well suited to small quantities of sample.

Fragility—By their nature, gravimetric apparatus are quite fragile and care must be exercised in their use. Volumetric equipment usually contains fragile glass components. Continuous flow instruments are all metal, except for the sample cell, and are the least fragile of the various apparatus.

Pressure Overshoot—The dosing requirement of gravimetric and volumetric apparatus can lead to pressure overshoot which may produce data off the isotherm in the hysteresis region. The continuous flow method is not susceptible to this phenomenon.

Adsorbates—Gravimetric instruments can accommodate any adsorbate. Continuous flow instruments are limited to adsorbates that will not condense at room temperature. Volumetric apparatus requires various internal volume and manometer alterations when adsorbates with low vapor pressures are used.

Relative Pressure—Saturated vapor pressures are determined independently of the adsorption method employed. The adsorbate pressure must be measured by manometer or gauges when gravimetric or volumetric apparatus are used. Barometric pressure must be known for continuous flow measurements.

Sensitivity—The continuous flow method can detect signals corresponding to less than 0.001 cm^3 of adsorption with 1% accuracy. This makes the thermal conductivity detector used in continuous flow instruments considerably more sensitive to small amounts of adsorption than the volumetric or gravimetric methods.

Permanent Record— Vacuum electronic microbalances and continuous flow instruments provide electronic signals for a permanent record.

Errors—Both the gravimetric and volumetric methods are subject to cumulative errors. The continuous flow method produces each data point independently and is not subject to error accumulation.

CHAPTER EIGHTEEN

Chemisorption

1. Introduction

In Chapter 3, the various forces responsible for physical adsorption were discussed. These include dispersion as well as coulombic forces. Chemical adsorption arises from the inability of surface atoms to interact symmetrically in the absence of a neighbor above the plane of the surface. For this reason, surface atoms often possess electrons or electron pairs that are available for bond formation.

Because chemisorption involves a chemical bond between the adsorbate and adsorbent, only a single layer of chemisorption can occur. Physical adsorption on top of the chemisorbed layer and diffusion of the chemisorbed layer beneath the surface can obscure the fact that chemisorbed material can be only one layer in depth.

Often the attempt is made to distinguish between physical and chemical adsorption on the basis of the heat of adsorption. However, this is not an entirely satisfactory procedure. The smallest physical heat of adsorption will be slightly greater than the heat of liquefaction of the adsorbate. Were this not true the vapor would condense and not be adsorbed. The upper limit for physical adsorption may be higher than 20 kcal $mole^{-1}$ for adsorption on adsorbates with very narrow pores such as silica gel or zeolites. The heat of chemisorption ranges from over 100 kcal $mole^{-1}$ to less than 20 kcal mole^{-1}. Therefore, only very high or very low heats of adsorption can be used as criteria for the type of adsorption process. A more definitive criterion of whether a particular interaction is physical or chemical is to search for reaction products. Other techniques, such as isotope exchange, electron diffraction, adsorption spectroscopy, magnetic susceptibility, and electron spin resonance are just a few of the modern methods that can be used to establish a detailed description of the adsorbate–adsorbent interaction.

2. Chemisorption Equilibrium and Kinetics

The measurement of the equilibrium between the gaseous and the chemisorbed state is frequently difficult because of the very low equilibrium pressures required to saturate the surface. Often, in the case of strong interactions the monolayer is completed at very low pressures, even less than one torr, as shown in Figure 18-1.

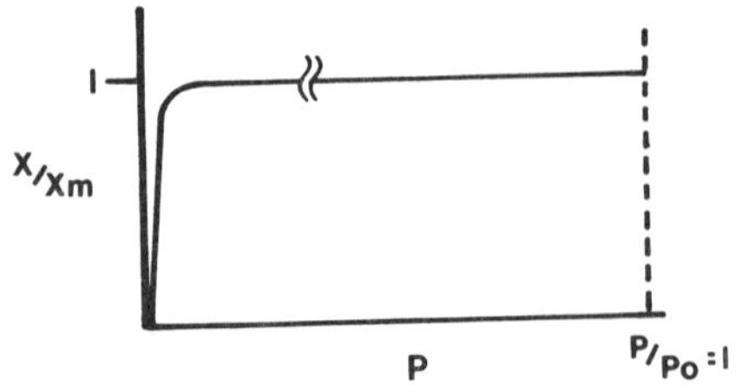

Figure 18-1 Type I chemisorption isotherm for a strongly bound adsorbate.

When the adsorbate is strongly bound and localized to one adsorption site, as in chemisorption, the Langmuir equation is applicable:

$$\frac{W}{W_m}=\frac{KP}{1+KP} \qquad \text{[cf. 5.11]}$$

At sufficiently low pressure $KP \ll 1$, the Langmuir equation becomes

$$\frac{W}{W_m} = KP \qquad [18.1]$$

This region of the isotherm is the Henry's law region; the uptake being directly proportional to the pressure. At high pressures, $KP \gg 1$, so that Langmuir's equation becomes $W = W_m$.

Figure 18-2 illustrates some of the essential differences between physical and chemical adsorption.[132]

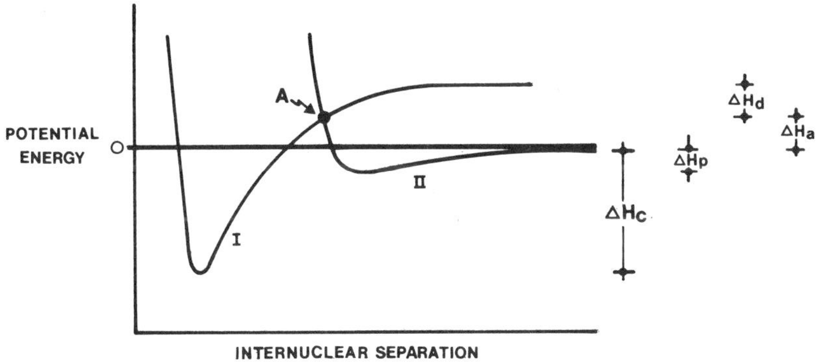

Figure 18-2 Potential energy curves for physical and chemisorption.

Curves I and II represent potential energy plots for chemical and physical adsorption, respectively. The zero of potential energy is taken, as is customary, at infinite separation of the interacting species. The minimum in curve I, below zero potential energy, is equal to the heat of chemisorption, ΔH_C and the minimum of curve II is equal to the heat of physical adsorption, ΔH_P. The fact that curve I lies above zero potential energy at large internuclear separations implies that the chemisorbed gas is in an activated state or has undergone dissociation. The term ΔH_D, thus, is the heat of activation or dissociation. If activation or dissociation does not occur, then curve I would approach zero potential energy asymptotically, similar to curve II. The minimum of curve I occurs at a smaller internuclear separation than that of curve II because chemical bonding, involving orbital overlap, will bring nuclei closer together than can the less energetic physical adsorption forces.

The transition from physical adsorption to chemisorption occurs at point *A*. The potential energy at *A* is in excess of that for the adsorbate and the adsorbent when separated and represents the activation energy required for chemisorption, ΔH_A. If curve I resided more to the right or curve II more to the left, then the transition from physical to chemical adsorption would occur with no activation energy since the crossover point would reside beneath zero potential energy.

The concept of activated adsorption explains why often the heat of adsorption is small at low temperatures and large at higher temperatures. This is because at low temperature the heat is due to physical adsorption whereas at higher temperature sufficient thermal energy is available to provide the required activation for chemisorption with its associated high heat of adsorption. Also, the rates of adsorption frequently decrease at higher temperatures indicating that the adsorption mechanism undergoes a transition to an activated process. A further frequent observation is that the amount adsorbed varies with temperature as shown in Figure 18-3.[132] The initial decrease is due to thermally desorbing the physically adsorbed gas. Subsequently, the quantity adsorbed increases with increasing temperature due to commencement of activated chemisorption. Finally, the curve slopes downward when sufficiently high temperature is reached to thermally desorb the chemisorbed state.

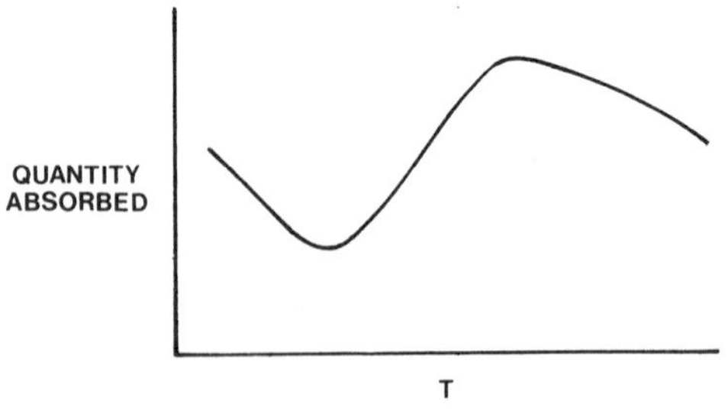

Figure 18-3 Variation in quantity adsorbed with temperature.

3. Determination of Chemisorption Isotherms

When the initial slope of a chemisorption isotherm is not large so that monolayer coverage is approached at conveniently measurable pressures, the gravimetric, volumetric, or continuous flow methods can be employed. The gravimetric and volumetric methods for determining this type of isotherm are employed in their usual manner provided that no reaction products contribute to the equilibrium pressure. The only essential difference when measuring chemical and physical adsorption using these techniques, is that the sample is maintained at an elevated temperature rather than in a coolant bath.

When using the continuous flow method, however, some additional versatility is available in chemisorption measurements. For example, when data is required at an adsorbate pressure of 0.1 atm, a 10% mixture of adsorbate mixed with an inert carrier gas is passed through the apparatus with the sample cooled to a temperature at which no chemisorption can occur. Upon warming the sample to the required temperature, adsorption occurs which produces an adsorbate-deficient peak that is calibrated by injecting carrier gas into the flow stream. Equation [14.9] is then used to calculate the quantity adsorbed. This process is repeated for each concentration required. Caution must be exercised to avoid physical adsorption when the sample is cooled to prevent chemisorption. Should this occur, the adsorption peak due to chemisorption can be obscured by the desorption peak of physically bound adsorbate when the sample is heated.

When there is no satisfactory temperature at which neither physical or chemical adsorption will occur, an alternate method can be used. Referring to Figure 14-9, valve V_6 can be rotated to shunt the gas flow past the cell. After flowing pure carrier gas through the sample cell and purging out any contaminants, the cell is isolated and the flow is switched, for example, to a 10% mixture. When a steady base line is obtained, valve V_6 is rotated to admit the mixture into the sample cell. Presuming that the sample is maintained at a temperature at which chemisorption will occur, a peak will be produced with an area proportional to the amount of chemisorption and to the volume of carrier gas swept out of the cell void volume. The cell is again isolated and the flow concentration increased, for example, to 20%. Again, when the flow is admitted into the sample cell, the area under the resulting peak will be proportional to the amount adsorbed plus the volume of 10% mixture swept out of the void volume. This process is continued until all the required data is obtained.

Because each signal must be corrected for the volume of the previous gas concentration contained in the cell, it becomes necessary to measure the cell void volume. This is accomplished by filling the cell, containing the sample, with an inert gas, then isolating the cell and flowing a second nonreacting gas through the system. When valve V_6 is rotated, the cell will be purged and the resulting peak is calibrated with a known volume of inert gas. The cell void volume V_v is given by

$$V_v = \frac{A}{A_{\text{cal}}} V_{\text{cal}} \qquad [18.2]$$

where A is the area produced when the cell is opened, thereby purging the trapped gas and A_{cal} is the area produced by injecting a known volume V_{cal} of the same gas. The volume adsorbed is then expressed as

$$V_{\text{ads}} = V_{\text{carrier}} \left(\frac{C_{\text{absorbate}}}{C_{\text{carrier}}}\right)\left(\frac{A_{\text{sig}}}{A_{\text{cal}}}\right) - V_v \qquad [18.3]$$

The terms in equation [18.3] are defined as follows:

V_{ads} is the volume chemisorbed, measured at room temperature and pressure.

V_{carrier} is the volume of carrier gas injected when calibrating the peak produced by opening and purging the sample cell.

$C_{\text{adsorbate}}$ and C_{carrier} are the adsorbate and carrier gas concentrations in the flow stream. These concentrations are most conveniently expressed as mole fractions.

A_{sig} is the area under the peak resulting from adsorption when the cell is opened and purged.

A_{cal} is the area under the calibration curve using pure adsorbate.

V_v is obtained from equation [18.2].

If two inert gases are not available an empty cell can be calibrated using pure adsorbate and carrier gas. The void volume can then be calculated as the difference between the empty cell volume and the sample volume, presuming that the sample's density is accurately known.

For maximum accuracy those portions of the void volume not heated should be maintained at constant temperature to prevent the void volume from varying during the analysis.

Often chemisorption produces reaction products that are swept out of the sample cell to the detector. In such cases, to prevent obscuring the adsorption signal, it is necessary to remove the reaction products from the flow stream. Depending upon their nature they can often be removed chemically or by condensation in a cold trap.

When hydrogen is used as the adsorbate, helium will not adequately serve as a carrier gas because these two gases exhibit similar thermal conductivities. In this case nitrogen or any other nonreacting gas can serve as the carrier.

4. Surface Titrations

When the adsorbate is strongly bound to the surface such that desorption does not occur or occurs at a negligibly slow rate, the monolayer capacity of the adsorbent can be obtained by titrating the surface with the adsorbate.[133–134] This method requires flowing an inert gas over the sample and then injecting a quantity of adsorbate into the flow stream. As the band of adsorbate is carried over the sample, some adsorption will occur. The detector will respond to the quantity of adsorbent which is not chemisorbed; subsequent adsorbate injections will produce successively larger peaks as the surface approaches saturation. The surface is saturated when two or more successive peaks exhibit the same signal area. The total volume adsorbed V_{ads} is given by

$$V_{ads} = V_1^{inj} - \frac{A_1}{A_1^{sat}} V_1^{inj} + V_2^{inj} - \frac{A_2}{A_2^{sat}} V_2^{inj} + \cdots + V_i^{inj} - \frac{A_i}{A_i^{sat}} V_i^{inj} \qquad [18.4]$$

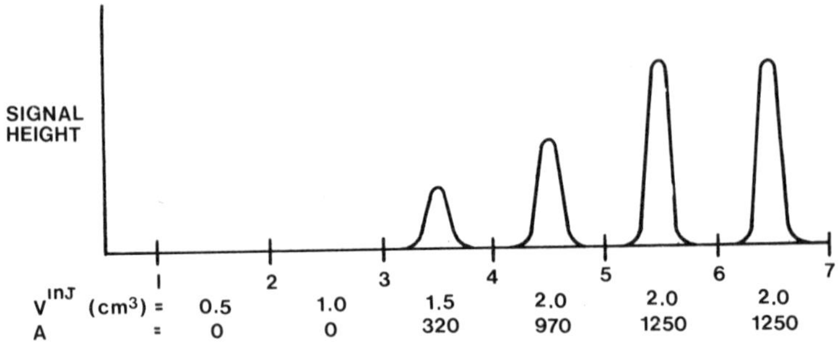

Figure 18-4 Chemisorption titration for the measurement of total uptake.

where V^{inj} is the volume of adsorbate injected into the flow stream, A is the signal area after the band of adsorbate is carried over the adsorbent, and A^{sat} is the signal area corresponding to the injected volumes when no adsorption occurs. Figure 18-4 illustrates the quantity of uptake for a hypothetical case. For maximum accuracy a calibration curve should be prepared to eliminate the detector nonlinearity as a source of error.

CHAPTER NINETEEN

Density Measurement

In many areas of powder technology the need to measure the powder volume or density often arises. For example, powder bed porosities, volume-specific surface area, sample cell void volumes, as well as numerous other calculated values all require accurately measured powder densities or specific volumes. It is appropriate, therefore, to introduce some discussion on powder density measurements.

1. True Density

The true density is defined as the ratio of mass to the volume occupied by that mass. Therefore, contribution to the volume made by pores or internal voids must be neglected when measuring the true density. Regardless of whether pores and internal voids are present, the density of fine powders is often not the same as that of larger pieces of the same material because in the process of preparing many powders, those atoms or molecules located near the surface are often forced out of their equilibrium position within the solid structure. On large pieces of material the percentage of atoms near the surface is negligibly small. However, as the particle size decreases, this percentage increases with its resultant effect upon the density.

If the powder has no porosity the true density can be measured by displacement of any fluid in which the solid remains inert. Usually, however, the solid particles contain pores, cracks, or crevices that will not be penetrated by a displaced liquid. In these instances the true density can be measured by using a gas as the displaced fluid. Helium is preferred for this measurement since it can enter the finest pores.

Apparatus used to measure solid volumes are often referred to as pyknometers or pycnometers after the Greek "pyknos" which means compaction. Once the sample volume and mass have been determined the density is calculated. Figure 19-1 is a schematic

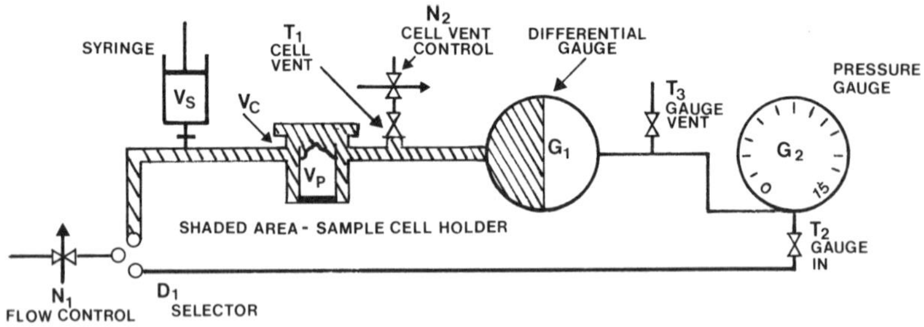

Figure 19-1 Null-pycnometer for true powder densities. Courtesy of Quantachrome Corporation.

diagram of a commercial pycnometer capable of making accurate powder volume measurements. The volume V_c within the shaded area is considered the sample cell holder. When the syringe is depressed, the pressure in the cell holder increases and the differential gauge G_1 indicates a pressure difference. Gas is then allowed to flow under pressure through needle valve N_1 and valves D_1 and T_2 to pressurize gauge G_2 while simultaneously rebalancing or nulling gauge G_1. When gauge G_1 is exactly nulled the reading on gauge G_2 is taken. This procedure is performed two times; once with the cell empty and again with powder in the cell. From the two readings taken on gauge G_2 the powder volume V_p can be calculated as

$$V_p = V_c\left(1 - \frac{P_2 - P_1}{P'_2 - P_1}\right) \qquad [19.1]$$

where

V_p = volume of powder (cm^3)
V_c = volume of sample cell holder (cm^3)
P_2 = reading on gauge G_2 when gauge G_1 is nulled with no powder in the sample cell
P'_2 = reading on gauge G_2 when gauge G_1 is nulled with powder in the sample cell
P_1 = reading on gauge G_2 when open to the atmosphere. This value is made to read zero by adjusting the gauge mechanical zero.

Equation [19.1] is derived as follows: Prior to depressing the syringe plunger the volume of gas in the sample cell holder V_c (shaded area in Figure 19-1) and the volume of gas in the syringe V_s are isolated. The gas equation describes the state of the system as

$$P_1(V_c + V_s) = n_1 RT \qquad [19.2]$$

where

P_1 = ambient pressure
V_c = volume of sample cell holder
V_s = volume of syringe
n_1 = moles of gas contained in the syringe and sample cell holder with no sample
R = gas constant
T = ambient absolute temperature

When the syringe piston is depressed, the pressure in the cell holder increases to a higher value as indicated by differential gauge G_1. The movement of the mechanism inside gauge G_1 causes the cell holder volume to change. Therefore, a flow of gas is directed to the opposite side of the differential gauge until it again reads zero. When this condition is achieved the pressure in gauge G_2 is P_2 and equation [19.2] becomes

$$P_2 V_c = n_1 RT \qquad [19.3]$$

Combining equations [19.2] and [19.3] and solving for the cell volume gives

$$V_c = \frac{P_1}{P_2 - P_1} V_s \qquad [19.4]$$

With a volume of powder V_p in the sample cell, equation [19.2] becomes

$$P_1(V_c + V_s - V_p) = n_2 RT \qquad [19.5]$$

and equation [19.3] becomes

$$P_2'(V_c - V_p) = n_2 RT \qquad [19.6]$$

Here, n_2 represents the moles of gas in the system when the powder sample is present. Equations [19.5] and [19.6] combine to yield

$$V_c - V_p = \left(\frac{P_1}{P_2' - P_1}\right) V_s \qquad [19.7]$$

Dividing equation [19.7] by equation [19.4] gives

$$V_p = V_c \left[1 - \frac{P_2 - P_1}{P_2' - P_1}\right] \qquad [\text{cf. } 19.1]$$

The preceding derivation shows that the calculated powder volume is independent of the syringe volume. It is necessary that the syringe introduce only a reproducible volume so that the V_s terms cancel when equation [19.7] is divided by equation [19.4].

By zeroing gauge G_2 prior to each measurement, the terms $(P_2 - P_1)$ and $(P_2' - P_1)$ are read as pressures above ambient or simply P_2 and P_2'. Thus equation [19.1] can be rewritten as

$$V_p = V_c \left(1 - \frac{P_2}{P_2'}\right) \qquad [19.8]$$

The volume of the sample cell holder is calibrated by replacing the powder with carefully machined brass or steel cylinders with accurately known volumes V_{cyl}. Equation [19.8] can be rewritten

$$V_c = \frac{V_{cyl}}{(1 - P_2/P_2')} \qquad [19.9]$$

to obtain the sample cell holder volume V_c.

When measuring powder volume in the manner described previously, it is necessary to avoid using any gas that can be even slightly adsorbed. If even one-thousandth of a monolayer were adsorbed, the equivalent volume of gas would be in the order of 0.001 cm^3 for each 2.84 m^2 of area if nitrogen were used. Since the sample cell used in the apparatus described in Figure 19-1 can hold 130 cm^3, the sample's total surface area can be hundreds or even thousands of square meters. Thus errors of 0.1–1.0 cm^3 can be incurred if the gas is even slightly adsorbed. For this reason helium is recommended in any gas pycnometer.

A second source of error encountered with high-area powders is the annulus volume that exists between the powder surface and the closest approach distance of a gas molecule. Assuming that the closest approach of the center of mass of the helium atom to the powder surface is 0.5 Å or 5×10^{-11} meter and that the specific area of the powder is 1000 $m^2\ g^{-1}$, there will exist an annulus volume of $5 \times 10^{-8}\ m^3\ g^{-1}$ or $5 \times 10^{-2}\ cm^3\ g^{-1}$. This can represent a density error of 5% on materials with densities near unity.

2. Apparent Density

When the fluid displaced by powder does not penetrate all the pores the density will be less than the true density. When measuring densities by liquid displacement an apparent density is obtained that can differ according to the liquid used. Therefore, when reporting apparent density, the liquid used should also be reported.

3. Bulk Density

The volume occupied by the solid plus the volume of voids when divided into the powder mass yields the bulk density. Therefore, when powder is poured into a vessel, say a graduated cylinder, the bulk density is the mass divided by the volume of the powder bed.

4. Tap Density

The tap density is another form of bulk density except that the container has been tapped or vibrated in a specified manner to achieve more efficient particle packing. The tap density is, therefore, usually greater than the bulk density.

5. Effective Density

A particle may contain embedded foreign bodies that may increase or decrease its density. It may also contain blind pores that are totally encapsulated by the particle and effectively reduce the particle's density.

CHAPTER TWENTY

Mercury Porosimetry

1. Introduction

Although many high-pressure mercury porosimeters have been constructed, they all have several essential components, which are perhaps different in their design but nevertheless are common to each apparatus. These components include the following:

1. Cell to hold the test sample
2. Dilatometer or stem attached to the sample cell in which the mercury level varies with intrusion or extrusion
3. Vacuum filling apparatus to remove trapped air from the pores and for transferring mercury into the sample cell
4. Pressure generator
5. High-pressure vessel to contain the sample cell
6. Probe to measure the mercury level
7. Hydraulic fluid to transmit the pressure to the dilatometer

2. Pressure Generators

Commercial mercury porosimeters achieve maximum pressures in the range 50,000–60,000 psi which corresponds to pore radii of

0.0018 μm assuming 140° for the wetting angle (cf. equation [11.5]). Although higher pressures can be achieved, the interpretation of data much above 60,000 psi becomes questionable in view of the uncertainty regarding the meaning of wetting angles and surface tension in pores not much wider than the diameter of a few mercury atoms.

Usually the required pressures are obtained with a reciprocating pressure intensifier which transmits a force over a wide area to a smaller area, thus increasing the applied pressure. Check valves are employed to direct the fluid flow into and out of the intensifier on each cycle. Because of the reciprocating operation of the pressure intensifier, most porosimeters operate on an incremental basis. That is, the intruded volumes are measured as discrete data points after each incremental change of pressure.

3. Dilatometer

A dilatometer is used to measure the intrusion volume and several methods are used to measure the change in mercury level within the dilatometer stem as intrusion takes place. The first method involves the use of platinum or platinum alloy resistance wire placed coaxially in the stem of the dilatometer tube. As the mercury level decreases, the amount of resistance wire exposed increases, thereby providing a voltage that increases linearly with decreasing mercury level. Very small current is used to minimize resistive heating of the mercury. Platinum or platinum–iridium alloy may be employed as the resistance wire because they do not amalgamate with mercury.

A second method for monitoring the change in mercury level is the use of a follower probe. As the level decreases, electrical contact between the probe and the mercury level is broken. This actuates a drive motor which moves the probe, through a pressure seal, until contact is reestablished. The drive motor angular displacement is proportional to the change in mercury level.

Figure 20-1 illustrates a third method of measuring the mercury level. The mercury in the dilatometer stem constitutes one plate of a capacitor and a metal sheath coaxially surrounding the dilato-

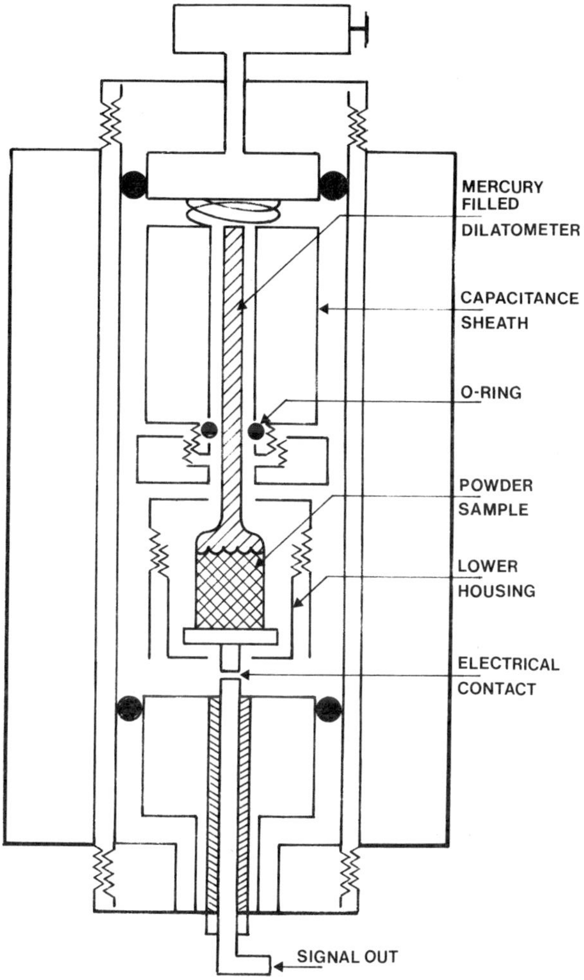

Figure 20-1 Capacitive probe; high pressure cavity and seals for continuous intrusion measurements.

meter serves as the second plate. The capacitance changes with the mercury level due to the variation of the effective plate area.

4. Continuous Intrusion

A recent development[115] is a mercury porosimeter capable of producing continuous plots of both the intrusion and extrusion curves. Figure 20-2 illustrates this apparatus. A pressure generator, driven by a 1 hp motor through a 50 to 1 gear reduction, uses a threaded shaft to drive a stainless steel rod into a narrow cavity filled with hydraulic oil. As the rod penetrates into the cavity, the oil is compressed in the entire apparatus. The total volume displaced by the rod is 20 cm^3 which is sufficient to produce pressure up to 60,000 psig on thc mercury column. The rate of pressurization can be controlled by the motor speed. A strain gauge transducer is located near the top of the generator and provides an electrical signal in proportion to the pressure.

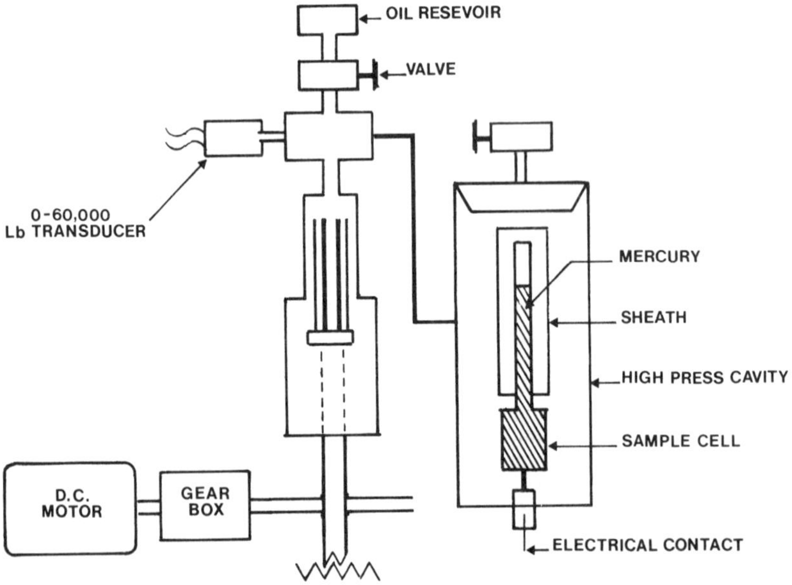

Figure 20-2 Continuous recording mercury porosimeter.

The pressure vessel containing the sample cell is connected to the pressure generator with high pressure tubing. An oil reservoir near the top of the apparatus feeds oil to the pressure vessel as required for filling. The pressure vessel has removal caps on the top and bottom for loading the sample cell and for easy draining in case a glass sample cell should break under test. The caps are screwed onto the pressure vessel and need be only finger tight to retain 60,000 psig.

The mercury level is monitored with a capacitance probe as shown in Figure 20-1. The signal from the capacitance probe and from the pressure transducer are plotted on an X–Y recorder as intrusion volume and pore radius, respectively.

Some of the advantages of this design include the following:

a. Continuous plots of both intrusion and extrusion curves are obtained.

b. At nominal drive speeds a complete analysis takes about 15 minutes.

c. At no time along the intrusion curve does the pressure relax. Therefore, no opportunity is provided for a small quantity of extrusion to occur following intrusion. This condition can produce a data point somewhere in the hysteresis region between the intrusion and extrusion curves.

d. The effects of mercury compression and the compressive heating of the hydraulic oil are electronically compensated. Therefore, the need to make blank runs is unnecessary for all but the most exacting analysis. Blank runs made on cells filled with mercury only show less than 1% of full scale signal over the entire operating range from 0 to 60,000 psig.

5. Logarithmic Signals

The signals produced by the strain gauge can be amplified linearly or logarithmically. When in the linear mode the displacement of the X–Y recorder along the X or pressure axis is inversely proportional to the pore radius as given by equation [11.4]

$$Pr = 2\gamma \cos \theta \qquad \text{[cf. 11.4]}$$

When the strain gauge signal is fed to the recorder as the logarithm of the pressure, the following relationships exist using A as $2\gamma \cos \theta$

$$\log P = \log A - \log r \qquad [20.1]$$

Since the electrical signal E is proportional to $\log P$ one can write

$$\log P = kE \qquad [20.2]$$

Combining equations [20.1] and [20.2] and taking derivatives of each side, yields

$$kdE = -\frac{dr}{r} \qquad [20.3]$$

Equation [20.3] is in the form of a first-order decay equation, which describes a first-order chemical reaction or radioactive decay. It establishes that the pore radius decreases logarithmically with increasing signal as described by equation [20.4], which is obtained by integrating equation [20.3]

$$r = Ke^{-kE} \qquad [20.4]$$

Therefore, when semilog paper is used in the X–Y recorder upside down so that the recorder pen sweeps from larger to smaller pore radii, the pore radius can be plotted directly.

6. Low-Pressure Intrusion

In order to measure pores larger than 7 μm it is necessary to monitor intrusion at pressures less than atmospheric. Figure 20-3 is a diagram of a filling mechanism that can be used for monitoring intrusion up to atmospheric pressure as well as outgassing of the sample. The filling mechanism has a provision for heating the sample to hasten the rate of outgassing.

After pumping to sufficiently low vacuum as indicated by the thermocouple gauge, usually about 50 millitorrs, the glass vacuum housing is rotated toward the vertical. Mercury flows into the narrow end of the housing and covers the end of the dilatometer stem to about one inch. The selector valve is rotated to admit air and the bleed rate is adjusted with the needle valve. When the

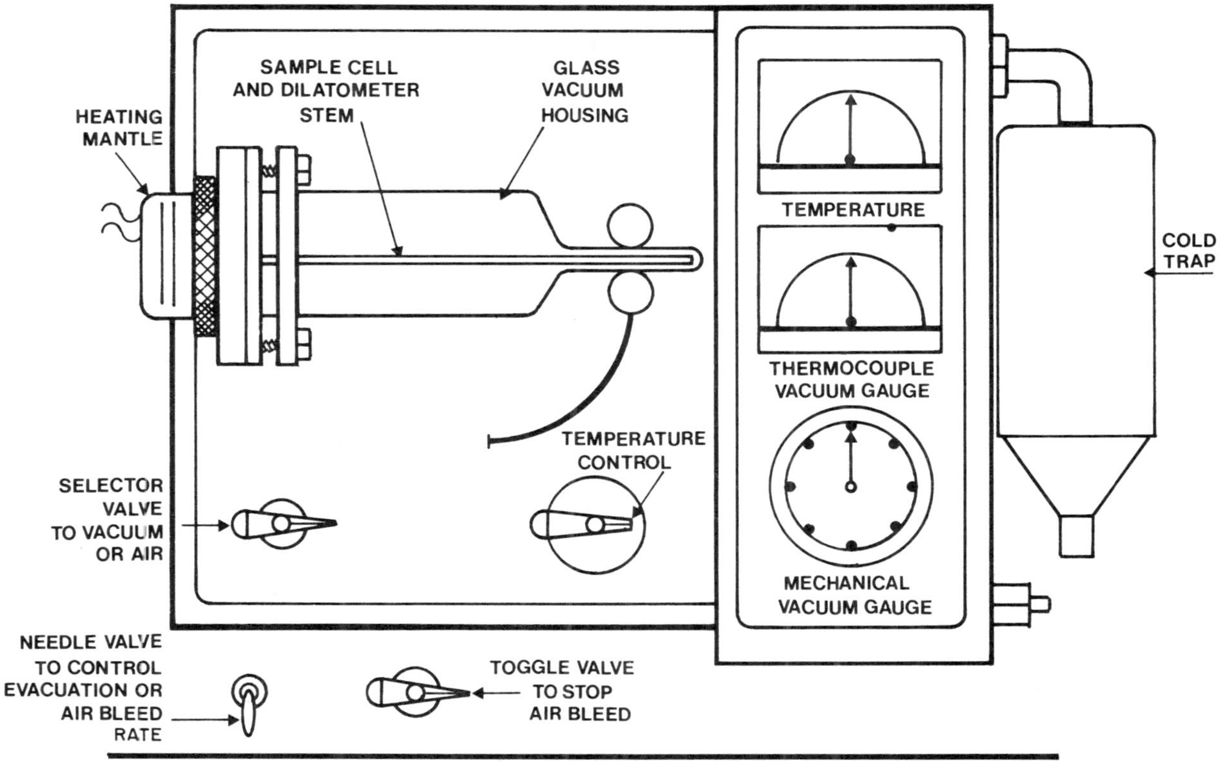

Figure 20-3 Filling mechanism for sample conditioning and low-pressure measurements.

sample cell and stem are filled with mercury the bleed is stopped, the housing returns to the horizontal position, and the pressure is read on the mechanical gauge. Thereafter small additions of air create a pressure increase which forces mercury to intrude into the large voids. The intrusion volume is obtained by measuring the displacement of mercury in the dilatometer stem. When the pressure in the vacuum housing reaches ambient, the sample cell is removed and inserted in the high pressure porosimeter for further intrusion volume measurements up to 60,000 psig.

7. Contact Angle for Mercury

A convenient method for measuring the wetting angle was developed by Poynting and Thompson.[135] When a small drop of mercury is placed on the surface of a smooth bed of powder it assumes a spherical shape except for the portion in contact with the surface. As additional mercury is added to the drop, the height increases until it reaches a maximum value. Further additions of mercury increase the drop diameter with no additional increase in the height. Figure 20-4 illustrates the change of drop shape as mercury is added.

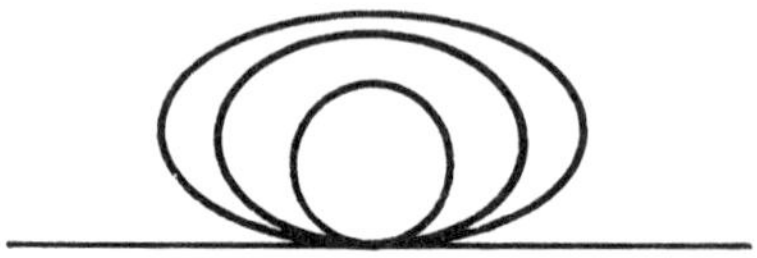

Figure 20-4 Change of mercury drop shape with size.

The relationship between angle θ and the maximum drop height $h_{\max}$ is

$$\cos\theta = 1 - \frac{\rho g h_{\max}^2}{2\gamma} \qquad [20.5]$$

where ρ and γ are the density and surface tension of mercury, respectively, and g is the acceleration of gravity.

Devices to accurately measure h_{max} are commercially available.[115] A highly accurate micrometer is employed to measure the maximum height of the mercury drop above a thin bed of packed powder.

Bibliography

1. Coulter Electronics, Hialeah, Florida.
2. C. Orr, Jr., and J. M. Dalla Valle, *Fine Particle Measurement*, Macmillan, New York, 1959.
3. T. Allen, *Particle Size Measurement*, Chapman and Hall, London, 1974.
4. F. London, *Z. Phys.*, **63**, 245 (1930).
5. S. Brunauer, L. S. Deming, W. S. Deming, and E. Teller, *J. Am. Chem. Soc.*, **62**, 1723 (1940).
6. I. Langmuir, *J. Am. Chem. Soc.*, **40**, 1368 (1918).
7. S. Brunauer, P. H. Emmett, and E. Teller, *J. Am. Chem. Soc.*, **60**, 309 (1938).
8. T. L. Hill, *J. Chem. Phys.*, **14**, 268 (1946).
9. S. Lowell, *Powder Technol.*, **12**, 291 (1975).
10. S. Brunauer, L. E. Copeland, and D. L. Kantro, *The Gas Solid Interface*, Vol. 1, Ch. 3, Dekker, New York, 1967.
11. S. Brunauer, *Solid Surfaces and the Gas Solid Interface*, Advances in Chemistry Series, No. 33, American Chemical Society, Washington, D.C., 1961.
12. D. S. MacIver and P. H. Emmett, *J. Am. Chem. Soc.*, **60**, 824 (1956).
13. G. Sandstede and E. Robins, *Chem. Ing. Tech.*, **32**, 413 (1960).
14. S. J. Gregg and K. S. W. Sing, *Adsorption Surface Area and Porosity*, Academic, New York, 1967.
15. P. H. Emmett and S. Brunauer, *J. Am. Chem. Soc.*, **59**, 1553 (1937).
16. A. V. Kiselev and Y. A. Eltekov, *International Congress on Surface Activity*, II, p. 228, Butterworth, London, 1957.
17. B. L. Harris and P. H. Emmett, *J. Phys. Chem.*, **53**, 811 (1949).
18. H. K. Livingston, *J. Colloid Sci.*, **4**, 447 (1949).

19. T. N. Rhodin, *J. Phys Chem.*, **57**, 1437 (1953).
20. A. A. Isirikyan and A. V. Kiselev, *J. Phys. Chem.*, **65**, 601 (1961); **66**, 210 (1962).
21. T. L. O'Connor and H. H. Uhlig, *J. Phys. Chem.*, **61**, 402 (1957).
22. A. C. Zettlemoyer, A. Chand, and E. Gamble, *J. Am. Chem. Soc.*, **72**, 2752 (1950).
23. R. T. Davis, T. W. DeWitt, and P. H. Emmett, *J. Phys. Chem.*, **51**, 1232 (1947).
24. P. H. Emmett and S. Brunauer, *J. Am. Chem. Soc.*, **59**, 1553 (1937).
25. S. Brunauer and P. H. Emmett, *J. Am. Chem. Soc.*, **59**, 2682 (1937).
26. P. Chenebault and A. Schurenkampfer, *J. Phys. Chem.*, **69**, 2300 (1965).
27. W. D. Harkins and G. Jura, *J. Am. Chem. Soc.*, **66**, 919 (1944).
28. I. Langmuir, *J. Am. Chem. Soc.*, **39**, 1848 (1917).
29. A. W. Adamson, *Physical Chemistry of Surfaces*, 2nd ed., Ch. 3, Interscience, New York, 1967.
30. J. W. Gibbs, *The Collected Works of J. W. Gibbs*, Vol. 1, Longmans Green, New York, 1931.
31. E. A. Guggenheim and N. K. Adam, *Proc. Roy. Soc. London*, **154A**, 608 (1936).
32. P. H. Emmett, *J. Am. Chem. Soc.*, **68**, 1784 (1946).
33. W. D. Harkins and G. Jura, *J. Am. Chem. Soc.*, **66**, 1362 (1944).
34. S. J. Gregg and K. S. W. Sing, *Adsorption Surface Area and Porosity*, p. 301, Academic, New York, 1967.
35. J. L. M. Poiseuille, *Inst. France Acad. Sci.*, **9**, 433 (1846).
36. H. P. G. Darcy, *Les Fontaines Publiques de la Ville de Dijon*, Victor De Lamont, 1856.
37. J. Koseny, *Ber. Wein Akad.*, **136A**, 271 (1927).
38. P. C. Carman, *J. Soc. Chem. Ind., London, Trans. Commun.*, **57**, 225 (1938).
39. P. J. Ridgen, *Road Res. Pap.*, No. **28**, (NMSO), 1954.
40. F. M. Lea and R. W. Nurse, *Trans. Inst. Chem. Eng.*, **25**, 47 (1947).
41. P. C. Carman, *Flow of Gases Through Porous Media*, Butterworth, London, 1956.
42. T. Allen, *Particle Size Measurements*, Ch. 15, Chapman and Hall, London, 1930.
43. W. T. Thomson, *Philos. Mag.*, **42**, 448 (1871).
44. A. Zsigmondy, *Z. Anorg. Chem.*, **71**, 356 (1911).
45. J. W. McBain, *J. Am. Chem. Soc.*, **57**, 699 (1935).

46. L. H. Cohan, *J. Am. Chem. Soc.*, **60**, 433 (1938).
47. A. G. Foster, *J. Chem. Soc.*, 1806 (1952).
48. J. C. P. Broekhoff and B. G. Linsen, *Physical and Chemical Aspects of Adsorbents and Catalysts*, Ch. 1, (B. G. Linsen, Ed.), Academic, New York, 1970.
49. J. H. de Boer, *The Structure and Properties of Porous Materials*, p. 68, Butterworth, London, 1958.
50. D. W. Broad and A. G. Foster, *J. Chem. Soc.*, 447 (1946).
51. A. G. Foster, *Proc. Roy Soc. London*, **147A**, 128 (1934).
52. D. W. McKee, *J. Phys. Chem.*, **63**, 1256 (1959).
53. M. J. Brown and A. G. Foster, *J. Chem. Soc.*, 1139 (1951).
54. L. Gurvitsh, *J. Phys. Chem. Soc. Russ.*, **47**, 805 (1915).
55. B. C. Lippens, B. G. Linsen, and J. H. de Boer, *J. Catal.*, **3**, 32 (1964).
56. C. G. Shull, *J. Am. Chem. Soc.*, **70**, 1405 (1958).
57. B. C. Lippens and J. H. de Boer, *J. Catal.*, **4**, 319 (1965).
58. C. Pierce, *J. Phys. Chem.*, **57**, 149 (1953).
59. M. R. Harris and K. S. W. Sing, *Chem. Ind.* (*London*), 11 (1959).
60. R. W. Cranston and F. A. Inkley, *Adv. Catal.*, **9**, 143 (1957).
61. G. D. Halsey, *J. Chem. Phys.*, **16**, 931 (1948).
62. C. Orr and J. M. Dalla Valle, *Fine Particle Measurement*, Ch. 10, Macmillan, New York, 1959.
63. E. P. Barrett, L. G. Joyner, and P. P. Halenda, *J. Am. Chem. Soc.*, **73**, 373 (1951).
64. A. Wheeler, *Catalyst Symposia*, June, 1945 and June, 1946, Gibson Island A.A.A.S. Conference.
65. C. Pierce and R. N. Smith, *J. Phys. Chem.*, **57**, 64 (1953).
66. P. C. Carman, *J. Phys. Chem.*, **57**, 56 (1953).
67. T. D. Oulton, *J. Phys. Colloid Chem.*, **52**, 1296 (1948).
68. B. F. Roberts, *A Procedure for Estimating Pore Volume and Area Distributions from Sorption Isotherms*, National Meeting of the American Chemical Society, Sept. 1963.
69. W. B. Innes, *Anal. Chem.*, **29**, 1069 (1957).
70. S. Brunauer, R. S. Mikhail and E. E. Boder, *J. Colloid Interface Sci.*, **24**, 451 (1967).
71. A. V. Kiselev, *Usp. Khim.*, **14**, 367 (1945).
72. M. M. Dubinin, *Zh. Fiz. Khim.*, **34**, 959 (1960).
73. Th. Van der Plas, *Physical and Chemical Aspects of Adsorbents and Catalysts*, Ch. 9., (B. G. Linsen, Ed.), Academic, New York, 1970.
74. M. Polanyi, *Verh. Dtsch. Phys. Ges.*, **16**, 1012 (1914).

75. M. M. Dubinen, *Chem. Rev.*, **60**, 235 (1960).
76. M. M. Dubinen and D. P. Timofeev, *Zh. Fiz. Khim.*, **22**, 133 (1948).
77. M. M. Dubinen and L. V. Radushkevich, *Dokl. Akad. Nauk SSSR*, **55**, 331 (1947).
78. K. M. Nikolayev and M. M. Dubinen, *Izv. Akad. Nauk SSSR, Otd. Tekn. Nauk.*, 1165 (1958).
79. M. G. Kaganer, *Zh. Fiz. Khim.*, **32**, 2209 (1959).
80. R. H. Mikhail, S. Brunauer, and E. E. Bodor, *J. Colloid Interface Sci.*, **26**, 45 (1968).
81. S. Brunauer, *J. Colloid Interface Sci.*, **39**, 435 (1972); **41**, 612 (1972).
82. J. H. de Boer, B. G. Linsen, T. van der Plas, Th. and G. J. Zondervan, *J. Catal.*, **4**, 649 (1965).
83. E. W. Washburn, *Proc. Nat. Acad. Sci.*, **7**, 115 (1921).
84. L. C. Ritter and R. L. Drake, *Ind. Eng. Chem., Anal. Ed.*, **782**, 17 (1945).
85. P. Zwietering, *The Structure and Properties of Porous Solids*, p. 287, Butterworth, London, 1958.
86. L. G. Joyner, E. P. Barrett, and R. Skold, *J. Am. Chem. Soc.*, **73**, 3158 (1951).
87. C. N. Cochran and L. A. Cosgrove, *J. Phys. Chem.*, **61**, 1417 (1957).
88. M. M. Dubinen, E. G. Vishnyakova, E. A. Zhukovskaya, E. A. Leontev, V. M. Lukyanovich, and A. I. Saraknov, *Russ. J. Phys. Chem.*, **34**, 959 (1960).
89. S. Glasstone, *Textbook of Physical Chemistry*, 2nd ed., p. 450, Van Nostrand, New York, 1946.
90. T. A. Hatton, *Powder Technol.*, **19**, 227 (1978).
91. J. de D. Lopez-Gonzales, F. G. Carpenter, and V. R. Deitz, *J. Res. Nat. Bur. Stand.*, **55**, 11 (1955).
92. C. Orr and J. M. Dalla Valle, *Fine Particle Measurement*, p. 176, Macmillan, New York, 1959.
93. P. A. Faeth, *Adsorption and Vacuum Technique*, Institute of Science and Technology, Univ. of Michigan, Ann Arbor, 1962.
94. P. H. Emmett, *A.S.T.M. Symposium on New Methods for Particle Size Determination in the Sub-Sieve Range*, p. 95, 1941.
95. M. R. Harris and K. S. W. Sing, *J. Appl. Chem.*, **5**, 223 (1955).
96. P. H. Emmett and S. Brunauer, *J. Am. Chem. Soc.*, **59**, 1553 (1937).
97. S. W. Benson, *The Foundation of Chemical Kinetics*, p. 66, McGraw-Hill, New York, 1960.
98. F. A. P. Maggs, *Research* (*London*), **6**, 513 (1953).
99. W. F. K. Wynne-Jones and H. Marsh, *Carbon*, **1**, 281 (1964).
100. F. A. P. Maggs, *Nature* (*London*), **169**, 259, 793 (1952).

101. P. Zwietering and D. W. van Krevelin, *Fuel*, **33**, 331 (1954).
102. R. A. Beebe, J. B. Beckwith, and J. M. Honig, *J. Am. Chem. Soc.*, **67**, 1554 (1945).
103. G. G. Litvan, *J. Phys. Chem.*, **76**, 2584 (1972).
104. T. Edmonds and J. P. Hobson, *J. Vac. Sci. Technol.*, **2**, 182 (1965).
105. A. J. Rosenberg, *J. Am. Chem. Soc.*, **78**, 2929 (1956).
106. W. V. Loebenstein and V. R. Deitz, *J. Res. Nat. Bur. Stand.*, **46**, 51 (1951).
107. J. H. de Boer, *The Dynamical Character of Adsorption*, p. 33, Oxford University Press, London, 1953.
108. F. M. Nelson and F. T. Eggertsen, *Anal. Chem.*, **30**, 1387 (1958).
109. S. Karp and S. Lowell, *Anal. Chem.*, **43**, 1910 (1971).
110. S. Karp, S. Lowell, and A. Mustaccuiolo, *Anal. Chem.*, **44**, 2395 (1972).
111. D. Kourilova and M. Krevel, *J. Chromatogr.*, **65**, 71 (1972).
112. S. Lowell and S. Karp, *Anal. Chem.*, **44**, 1706 (1972).
113. S. Lowell, *Anal. Chem.*, **45**, 8 (1973).
114. S. W. Benson, *The Foundation of Chemical Kinetics*, p. 188, McGraw-Hill, New York, 1960.
115. Quantachrome Corp., Syosset, New York.
116. A. J. Haley, *J. Appl. Chem.*, **13**, 392 (1963).
117. B. D. Semonian and M. Manes, *Anal. Chem.*, **49**, 991 (1977).
118. J. N. Wilson, *J. Am. Chem. Soc.*, **62**, 1583 (1940).
119. E. Glueckauf, *Proc. Roy. Soc. London*, **186A**, 35 (1946).
120. E. Glueckauf, *J. Chem. Soc.*, 1302, 1308, 1315, 1327 (1947).
121. E. Glueckauf, *Discuss. Faraday Soc.*, **7**, 12 (1949).
122. D. De Vault, *J. Am. Chem. Soc.*, **65**, 532 (1943).
123. J. Weiss, *J. Chem. Soc.*, 297 (1943).
124. A. C. Offord and J. Weiss, *Nature* (*London*), **155**, 725 (1945).
125. R. Stock, Ph. D. Thesis, London University, London, 1955.
126. S. J. Gregg, *The Surface Chemistry of Solids*, Reinhold, New York, 1961.
127. E. E. Cremer and H. F. Huber, (N. Breuner, J. E. Callen, and M. D. Weiss, Eds.), p. 169, Academic, New York, 1962.
128. H. Malamud, H. Geisman, and S. Lowell, *Anal. Chem.*, **39**, 1468 (1967).
129. L. Cahn and H. R. Schutz, *Vac. Microbalance Tech.*, **3**, 29 (1962).
130. J. McBain and A. M. Bakr, *J. Am. Chem. Soc.*, **48**, 690 (1926).
131. D. M. Young and A. D. Crowell, *Physical Adsorption of Gases*, Butterworth, London, 1962.

132. S. Glasstone, K. J. Laidler, and H. Eyring, *The Theory of Rate Processes*, McGraw-Hill, New York, 1941.
133. J. F. Rhodes and S. Katz, Meeting of the American Ceramic Society, 1974.
134. H. L. Gruber, *Anal. Chem.*, **34**, 1828 (1962).
135. L. I. Osipow, *Surface Chemistry*, p. 233, Reinhold, New York, 1964.

Index